Athena Unbound

The Advancement of Women in Science and Technology

Why are there so few women scientists? Persisting differences between women's and men's experiences in science make this question as relevant today as it ever was. This book sets out to answer this question, and to propose solutions for the future.

Based on extensive research, it emphasizes that science is an intensely social activity. Despite the scientific ethos of universalism and inclusion, scientists and their institutions are not immune to the prejudices of society as a whole. By presenting women's experiences at all key career stages – from childhood to retirement – the authors reveal the hidden barriers, subtle exclusions and unwritten rules of the scientific workplace, and the effects, both professional and personal, that these have on the female scientist.

This important book should be read by all scientists – both male and female – and sociologists, as well as women thinking of embarking on a scientific career.

HENRY ETZKOWITZ is Director of the Science Policy Institute and Associate Professor of Sociology at the State University of New York at Purchase.

CAROL KEMELGOR is a psychotherapist and psychoanalyst in private practice in Westchester County, New York, and Director of the Center for Women in Science, at the Science Policy Institute, State University of New York at Purchase.

BRIAN UZZI is Associate Professor of Business and Sociology at the Kellogg Graduate School of Management at Northwestern University.

To my mother

MARY MIRIAM LIFSHITZ ETZKOWITZ

BA Hunter College 1933

Magna Cum Laude, Geology

H.E.

For LARRY

C.K.

ATHENA UNBOUND

THE ADVANCEMENT OF WOMEN
IN SCIENCE AND TECHNOLOGY

HENRY ETZKOWITZ
Science Policy Institute,
State University of New York at Purchase

CAROL KEMELGOR
Science Policy Institute,
State University of New York at Purchase

BRIAN UZZI
Kellogg School, Northwestern University

With:

MICHAEL NEUSCHATZ, *American Institute of Physics*
ELAINE SEYMOUR, *University of Colorado*
LYNN MULKEY, *University of South Carolina*
JOSEPH ALONZO, *Rockefeller University*

CAMBRIDGE
UNIVERSITY PRESS

PUBLISHED BY THE PRESS SYNDICATE OF THE UNIVERSITY OF CAMBRIDGE
The Pitt Building, Trumpington Street, Cambridge, United Kingdom

CAMBRIDGE UNIVERSITY PRESS
The Edinburgh Building, Cambridge CB2 2RU, UK
40 West 20th Street, New York, NY 10011–4211, USA
10 Stamford Road, Oakleigh, VIC 3166, Australia
Ruiz de Alarcón 13, 28014 Madrid, Spain
Dock House, The Waterfront, Cape Town 8001, South Africa

© Henry Etzkowitz 2000

http://www.cambridge.org

First published 2000

Printed in the United Kingdom at the University Press, Cambridge

Typeface Trump Mediaeval 9.5/15pt System

A catalogue record for this book is available from the British Library

Library of Congress Cataloguing in Publication data
Etzkowitz, Henry, 1940–
    Athena unbound: the advancement of women in science and
technology / Henry Etzkowitz, Carol Kemelgor, Brian Uzzi, with Michael
Neuschatz . . . [et al.].
        p. cm.
    ISBN 0 521 56380 1   ISBN 0 521 78738 6 (pbk)
        1. Women in science. 2. Women in technology. I. Kemelgor, Carol,
1944–II. Uzzi, Brian, 1960–III. Title.

Q130.E85 2000 500'.82–dc21 00–020997

ISBN 0 521 56380 1 hardback
ISBN 0 521 78738 6 paperback

Contents

Acknowledgements

We express our appreciation to the National Science Foundation and the Sloan Foundation for research support.

Introduction
Women in science: Why so few?

Why are there still so few women scientists, especially at the upper levels of the scientific professions? Persisting differences between women's and men's experience in science make this question as relevant today as when sociologist Alice Rossi posed it more than three decades ago at a conference on women in science at the Massachusetts Institute of Technology (Rossi, 1965).

The years since Rossi's groundbreaking analysis have witnessed the revival of the feminist movement and the increased entry of women into many professions. Women have become lawyers and doctors in significant numbers, albeit unevenly distributed into high and low status subfields of these professions. Despite significant advances, there is a continuing disproportionate lack of women in most scientific and engineering disciplines, especially at the upper reaches of the professions.

One such scientist, Leslie Barber, a female Ph.D. in molecular biology, decided to end her career as a research scientist shortly after being awarded the doctorate. She reflected upon the mixed experience of her male and female peers in a recent article (Barber, 1995). On the positive side, she found widespread evidence of encouragement for girls and women to pursue scientific professions from the media and from parents and teachers.

On the negative side, in comparing the career trajectories of the ten members of her graduate research group, equally divided into five men and five women, Barber noted significant differences. Whether or not the men had done well in their graduate careers, they had forged ahead in their professional lives. Among the women, three 'have left research altogether, while the other two languish in post-doctoral positions, apparently unable to settle on a next step.' Barber was initially surprised that, despite the unique story that each woman offered to explain her situation, the traditional pattern of relative exclusion of females from the scientific professions had been reproduced in her graduate cohort.

A guarded professional prognosis for both men and women could well be advised for a field such as physics, where the potential numbers of qualified applicants, vastly overwhelm traditional occupational demand (Linowitz, 1996). Certainly there has been a shift away from nuclear weapons and power plants, as well as from 'big science' projects such as the cancelled Superconducting Super Collider, which once gave virtually automatic multiple choices of employment to Ph.D. physicists. Although not unemployed, young physicists can often be found utilizing their quantitative and analytical skills in the back rooms of Wall Street or even in their own financial firms.

But how can the male–female divide in following scientific research careers, as identified by Barber, be explained for molecular biology, given the proliferation of biotechnology firms with research positions in recent years? Why has the increase in women entering graduate school not been fully translated into female scientists occupying higher positions in the field? Why has science lagged other professions in its inclusion of women? The answers to these questions, and the responsibility for repairing a less than optimal outcome, can be found primarily within science and secondarily in the larger society (National Research Council, 1940; Fox, 1994).

A LIFE COURSE ANALYSIS OF WOMEN IN SCIENCE

The thesis of this book is that women face a special series of gender related barriers to entry and success in scientific careers that persist, despite recent advances. Indeed, while some of their male contemporaries view female scientists as 'honorary men', others see them as 'flawed women' for attempting to participate in a traditional male realm (Longino, 1987; Stolte-Heiskanen, 1987; Barinaga, 1993).

Female scientists have been at odds over how to respond to these invidious distinctions. Should they insist that as scientists they are not different from men? On the other hand, given that science has historically been a male-dominated profession, should not women claim that they must have their needs taken into account in how the field is organized?

We focus the greater part of this book on the quality of women's experience in academic science, on the grounds that the university serves as a gateway into the larger scientific community. Our analysis is based on extensive systematic fieldwork that focuses both on the personal accounts of female and male graduate students and faculty members, and on the statistical analysis of aggregate demographic data and survey data on person-to-person ties in departments. In interviews with us, they discussed their experience in research groups and departments as well as their interaction with male and female peers and mentors.

Athena Unbound provides a life-course analysis of women in science from early childhood interest, through university, graduate school and the academic workplace. The book is based on several studies: (1) fifty in-depth interviews with female graduate students and faculty members in five science and engineering disciplines at two universities; (2) four hundred in-depth interviews and focus groups with female and male graduate students and faculty members in five science and engineering disciplines at eleven universities; (3) follow-up interviews with a sub-sample of graduate students and post-doctoral fellows interviewed in the previous study; (4) a quantitative survey of female graduate students and faculty members in five science and engineering disciplines at one university, focusing on publication experiences; and (5) interviews with very young children on their image of the scientist as a gender-related role.

In the following chapter we will begin to address the question raised in this introduction: why so few women in science? We will present quantitative evidence documenting how women's entry into and leakage from the ranks of graduate school education and university departments differ from men's. As society becomes more knowledge-intensive, ending any exclusion of women from science and technology becomes more pressing.

1 The science career pipeline

In this chapter we discuss the 'pipeline' thesis for improving women's participation in science. This 'supply side' approach assumes that if sufficient women are encouraged to enter the scientific and engineering professions, the gender gap in science and technology will disappear.

The scientific career track, from elementary school to initial employment, has been depicted as a 'pipeline' like those for the transport of fluids and gases such as water, oil or natural gas. The rate of flow into scientific careers is measured by passage through transition points in the pipeline such as graduation and continuation to the next educational level.

Nevertheless, the flow of women into science is through, 'a pipe with leaks at every joint along its span, a pipe that begins with a high-pressure surge of young women at the source – a roiling Amazon of smart graduate students – and ends at the spigot with a trickle of women prominent enough to be deans or department heads at major universities or to win such honours as membership in the National Academy of Science or even, heaven forfend, the Nobel Prize' (Angier, 1995). Even this negative depiction of the pipeline as a leaky vessel is too optimistic. As we shall see, many women are discouraged from pursuing their scientific interests far earlier in their educational career than graduate training.

Although the rate of women entering scientific professions has improved significantly, especially in the biological sciences, the numbers reaching high-level positions are much smaller than expected. In the United States, for example, decades after the science-based profession of medicine experienced a significant increase in female medical students (currently about 40% are women), only 3% of medical school deans and 5% of department heads are women. Dr Eleanor Shore, dean for faculty affairs at Harvard Medical School, recalled, 'Originally we thought if we got enough women in, the problem would take care of itself' (Angier, 1995). But it obviously has not.

Significant numbers of women enter the 'pipeline' and then leave at disproportionate rates, or function less effectively, as covert resistance to their participation creates difficulties. At best, the picture of women's participation in science in recent decades is mixed. Indeed, the pipeline analogy is unintentionally appropriate as an implicit criticism of the way that the recruitment to science takes place.

In addition to the positive meaning of steady flow and assured delivery, a pipeline also connotes a narrow, constricted vessel with few if any alternative ways of passage through the channel. At each age grade, the entry ways for women become narrower and increasingly restrictive. As more are excluded, the talent pool for the next level to draw upon becomes smaller.

Although the genders are almost equally represented in the early stages of the pipeline they increasingly diverge at the later stages, resulting in a much smaller proportion of women than men emerging from the pipeline. At the point of career choice, many women are diverted from the academic and research tracks, even though some who are trained as scientists pursue science-related careers such as scientific writing or administration. The U.S. science pipeline runs through a distinctly different educational landscape than its counterparts in many other countries, and it is worth taking a moment to describe the system here.

THE U.S. EDUCATIONAL SYSTEM

In contrast to most European, Latin American and other countries where a specialized course of study on one or a few related areas makes up virtually the entire undergraduate curriculum, the U.S. educational system does not expect students to make an early choice of careers. Even though an increasing number of secondary and even middle schools have occupational themes such as healthcare, art and science, all offer a general education. The flexibility of the U.S. undergraduate degree allows room for secondary education to remain unspecialized.

Students typically graduate from high school after twelve years of primary and secondary education at the age of seventeen or eighteen. Where to go to college or university becomes a serious issue in the third year of high school, although student and parental anxieties about getting into a prestigious college or university have pushed these concerns ever earlier. Again in contrast to countries with national systems of examinations at secondary school leaving, the U.S. high school offers an education that can vary widely in quality among schools and even within the same school. High school is still the quintessential U.S. social scene depicted in television programs and motion pictures of a youth culture focused on peer status, looks and athletic ability. Intellectual merit is not a leading status distinction except in a very few leading public and private high schools.

Universities also vary widely in quality of education and prestige, in contrast to Europe where university-level institutions are, more or less, expected to be on the same level. There is also a tradition in the U.S. of students going to university away from home, if it can be afforded. This makes the college decision a major turning point in life. It also marks the entry of the student into a nationwide educational and prestige gradation market. To take account of the wide differences in quality among secondary schools, an external system of exams offered by a non-profit corporation rather than a government agency was established in order to help universities sort potential students from a wide variety of backgrounds. Once university intake broadened from a select set of students attending college preparatory public and private high schools, as had been the case in the 1920s, to a mass education system, uniform measures were needed and the College Board examinations were established for this purpose.

The College Board examinations focus on general abilities in mathematical and analytical reasoning and are not directly tied to the high school curriculum. Therefore, a separate educational industry has grown up offering courses and tutoring to prepare students for these examinations, whose sponsors persist in insisting that formal preparation will do no good. Through these exams, high school grades, recommendations and sometimes an essay to be written on 'life goals', 'the most influential book I have read' or some such topic, combined with interviews by an alumnus or a college admissions officer, an initial selection is made.

High school graduates are sorted into more than 3,000 institutions of higher education, ranging from four-year baccalaureate colleges to universities offering Ph.D. degrees. However, this selection is still malleable since college students increasingly take time off from their studies to travel or work for a while and then decide to apply for transfer.

Almost 70% of U.S. high school graduates now continue on to post-secondary education. This is still in sharp contrast to the U.K. which has only in the past decade seen a rise from 10% to 30%, with an expected rise to 40% of secondary school leavers continuing on to university during the next decade.

In the U.S., general education continues from high school into the university. 'Distribution requirements' insure that students take one or more courses in the various spheres of knowledge such as science, art, history, languages and mathematics. In addition, many colleges and universities require students to take certain courses, typically in writing and the history of western civilization, as part of a general education program. In other countries such broad knowledge and skills are expected to be acquired in secondary schools, leaving the university career completely to specialized and professional training.

In the U.S. specialization begins at the baccalaureate level with declaring a major. 'A major' is a group of related courses in a disciplinary area such as history or biology, although it can also be an interdisciplinary group of courses in an area such as biology and society. An individual course typically consists of a sixteen-week series of class meetings totalling around three hours per week. It may combine lectures, class discussion and laboratories. Evaluation is likely to be some combination of laboratory exercises, short examinations or quizzes, a mid-term examination and/or a final examination. A research paper may also be required.

The course is the basic building block of undergraduate education and the credits attached to it, typically three or four, are added up to the requisite 120 for the degree with the major representing perhaps a third of that total. The European model would instead be the degree course with a set of requirements, lectures and examinations geared to measuring an end result rather than discrete pieces along the way, through the course.

The science major in the U.S. follows an intermediate format between the general U.S. undergraduate and specialized European educational models. Its courses typically must be taken in sequence and a larger proportion of the student's time is required. This leaves less time for electives, those courses apart form major, distribution or general education requirements in which students may follow a non-degree interest or simply take a course that has a reputation for being interesting, easy or challenging, whatever meets their needs!

Vocational choices can be put off at least until the second year of a four-year undergraduate career, or even later, unless one is in the sciences. Even if a science or engineering major is chosen late in the undergraduate career, courses can be made up in summer school or by taking an extra year for the degree. Some universities even offer a post-baccalaureate year program to prepare humanities and social science majors who have decided after graduation that they wish to go to medical school, a post-bachelor's degree program in the U.S. A year of chemistry, biology, physics and other related courses allows them to meet the basic requirements for admission.

The U.S. undergraduate model of education, based on courses, continues on into graduate school. A Ph.D. program typically begins with a set of courses during the first and second years whose purpose is to bring everyone up to the same level of basic knowledge in the field. Now, at this late stage, the U.S. system finally begins to follow the

European model, by evaluating students through an extensive 'qualifying' examination, cross-cutting an entire field.

Indeed, students do not necessarily have to prepare for the qualifying exam, the prerequisite for beginning research for the Ph.D. dissertation, by taking a set of courses. They may also study on their own, using reading lists, or more likely, in small groups of fellow students, so-called study groups, where old exams and problems are discussed. Again, this organized system of preparation for research is in contrast to the traditional European model in which a student tackles a research problem from the outset of the advanced degree process. There, the problem is often set in advance and candidates are advertised for in the scientific press.

Although the U.S. secondary and undergraduate education varies greatly in quality, it is at the graduate level that the U.S. excels. Research groups of a professor with graduate, undergraduate students and technicians are the basic building block of U.S. academic science. Assistant professors in the U.S., who would be junior researchers under a professor in many European countries, have the responsibility for raising their own research funds through competitive grants to start their own group. Success or failure in convincing the research community to fund their proposal is the prerequisite for attaining a permanent position in a U.S. research university. However, as we shall see, women and men experience the various stages and phases of this system quite differently.

THE LOSS OF WOMEN TO SCIENCE

With this system of education in mind, we return to the 'pipeline' hypothesis. This optimistic hypothesis has been at least partially disconfirmed by the mixed experience of the most recent generation of women in science and engineering. True, a large number of women in the U.S. major in science and engineering and a significant percentage of women receive BA degrees. As a result, the proportion of science and engineering bachelors' degrees going to women has almost doubled in three decades, rising from 25% in 1966 to 47% in 1995 (NSF, 1998:

171). But the number of women enrolled in graduate school is still significantly lower, at 38% in 1995 (ibid.:226–7). And the percentage who emerge with a Ph.D. in these disciplines is lower still, reaching only 31% by 1995. Even this figure is misleading, however, since it conceals sharp contrasts by discipline. Most of the progress is attributable to the greater presence of women in the life and social sciences, in contrast to the physical sciences and engineering. Highly unequal participation is still the norm in many fields.

These contrasts are, not surprisingly, most evident at the highest academic levels. Starting from what was then a relatively strong base of 16% in the 1960s, women increased their representation among Ph.D. biologists to 40% by the 1990s (see Table 1.1). From a smaller base of 7%, chemistry has seen a corresponding rise to 27%, while the geosciences increased more dramatically from 3% to 22% in the same period. However, although mathematics, physics and engineering have also seen substantial gains in the presence of women among doctorates, in none of these fields did the 1995 figure even reach one in five. Thus, starting from bases of 5% in mathematics, 2% in physics, and less than 1% in engineering in the 1960s, the proportion of Ph.D.s going to women has risen to 19%, 12% and 11% in the 1990s.

Table 1.1 *Women's share of science and engineering Ph.D.s, 1966–1997*

	1960s	1970s	1980s	1990s
Biology	16	21	33	40
Chemistry	7	10	19	27
Geosciences	3	6	16	22
Mathematics	5	10	15	19
Physics	2	4	8	12
All engineering	<1	1	6	11

Source: U.S. National Science Foundation, Survey of Earned Doctorates.

EUROPEAN COMPARISONS

Most European countries have shown similar patterns to the U.S. For example, in the United Kingdom, the starting point was so low in most fields that, even after some progress, women remain far below parity. In the late 1980s, female chemists in the U.K. were 35% of undergraduates, 24% of graduate students, 22% of post-doctoral researchers, 5.5% of lecturers, 1.5% of senior lecturers, 1% of the readers and 0% of professors. In U.K. academic science as a whole only 3% of professors and department heads were female, compared with 10% in the U.S. In France, there is a decreasing proportion of women physicists at the higher levels of government-sponsored research institutes (CNRS). At the lower levels, 42% of the best-qualified research assistants are women, perhaps in part reflecting their disproportionately low (16.8%) representation (Couture-Cherki, 1976).

The paucity of women in high-level scientific positions in the U.K. is exemplified by a footnote identifying the author of a preface to a volume on the condition of women in science: it notes that Professor Jackson was the first and only female professor of physics in the United Kingdom (Haas and Perucci, 1986). She is now deceased, but there were two female physics professors in British universities in the early 1990s (Healey, 1992). Nevertheless, the continuing low participation at higher career levels is a virtually universal cross-national phenomenon despite a history of improvement at the lower levels. University College London is a bright spot. The proportion of female professors at 9% is three times the national average. This is due to 'attention to problems of family and childcare'. Despite the bleakness of the overall situation, this instance demonstrates that actions can be taken that will significantly improve matters.

THE FALLACY OF THE 'SUPPLY SIDE'

The expectation that the problem of participation of women and minorities in the scientific and engineering professions could be solved with a bit of 'pump priming' is a supply side thesis. The supply side approach is codified in the so-called 'pipeline' thesis that recruiting more women is a sufficient strategy. By encouraging girls to study science, so the theory goes, participation of women and men in science will become more equal. Once this is accomplished, it is expected that one can then wait patiently for the next generation to witness women's inevitable rise to leadership positions in science in equal proportions to male scientists. Such a focus tends to neglect analysis of the 'demand side', especially organizational resistance to change and the persistence of barriers to entry of women into the scientific and engineering professions. Although there has been some recent progress, women continue to be chronically underrepresented in scientific careers, and their participation declines as one moves higher up the career ladder (Zuckerman, Cole and Bruer, 1991; National Research Council, 1993).

Role models

Some proponents of women in science believe that presenting young women entering the scientific and engineering professions with a picture of the resistance they will encounter will discourage them from going on. They believe that introducing young women to successful role models is the best way to enhance their chances of success.

A recent event hosted by the Section of Women In Science at the New York Academy of Sciences further illustrates the contradiction of celebrating the achievement of successful female scientists as an encouragement to girls to do science, rather than warning them about (and thus preparing them to meet) the possible obstacles. Several leading female scientists and engineers including Sheila Widnall, then Secretary of the Air Force, presented an account of their careers to an audience primarily composed of secondary school women, pursuing Westinghouse and other awards. Although one woman mentioned significant obstacles in her path, such as being turned down for tenure despite considerable research achievements, the overall tone of the meeting was upbeat and celebratory. The darker side of the scientific endeavor for females was played down.

As minorities move up educational and job ladders, it is expected that the problem of exclusion will be solved. However, a significant increase in women in academic science is unlikely to be realized simply by increasing the numbers of women who embark on a scientific career. Encouraging more women to enter the pipeline is at best a partial answer if so few are willing or able to come out at the other end and carry on professional careers in science.

2 Women and science: Athena Bound

Athena, the Greek mythological figure with strong female and male elements in her identity, personifies the dilemma of the contemporary female scientist. Contemporary female scientists are expected, and often expect themselves, to combine a demanding personal and professional life, without its effects on either. Even as some female scientists struggle to balance their professional and personal lives, others continue or are constrained to comply with a traditional 'male model' that rigidly subordinates the personal to the professional. Women in science comprise a diverse set of persons who, despite a common gender, do not embrace a collective identity.

Many successful women in scientific and engineering professions expect to have crossed a threshold into a work life in which gender is irrelevant. These fortunate few females are taken on as apprentices and, encouraged by their undergraduate professors, enter graduate school in the sciences and engineering. There again, they encounter an opaque competitive system that typically depletes their self-confidence.

Those women who complete the Ph.D. face a series of career choices that often needlessly clash with personal aspirations. As Athena found in pursuing her adventures as a woman in a higher world dominated by a male ethos, gender matters.

Alternate competing theses have been suggested to explain the resistance to women in science. It is not 'either/or'. Rather than 'barriers to entry', visible and invisible impediments to women pursuing a scientific career, or a 'glass ceiling' that places limits on recognition of achievement, difficulties exist at all stages and phases of the scientific career line.

Women who have avoided discouraging experiences at an earlier stage often encounter them later. For example, because women are often excluded from information and informal channels in graduate school, they have less access to 'social capital,' the network of relationships and connections, than their male peers. Without this network of professional and social psychological partners, women of equal or better 'human capital' (their skills and knowledge) are more likely to drop out of graduate school, and those who receive a Ph.D. lack the 'halo effect' that comes from inclusion in such a network.

When a relatively small number of women traverse the pipeline to win a faculty appointment the story is said to have ended successfully. Yet even at this juncture many highly effective women suddenly find themselves subtly ostracized while paradoxically expected to be 'role models' during the precarious tenure process. We call all of these disjunctures aspects of the 'cascade effect' in which the steady flow of energy can be short-circuited at any point, regardless of the level of achievement.

The experiences of women scientists begin and end with the consequences of social exclusion in an activity that necessitates, perhaps demands, community. All too often the consequences of social isolation and aloneness have been attributed to inherent deficits within the women themselves. The argument has been that they lack the right human capital for physically demanding and mathematically intensive scientific work, whether by nature's wisdom which has divided the gene pool or by self-selection into softer fields that permit greater attention to family. However, the experience of separateness and stigma makes more understandable the tendencies for self-blame, lack of self-confidence, fear of risk-taking and role confusion at the highest faculty level. These constraints on women arise from the way that society tracks and awards women and men differently, and are then manifested and reinforced at the organizational level (universities and departments) through discriminatory practices, misperceptions, and social networks that can include or isolate women.

Female scientists sometimes respond to the strictures against them by adopting a research strategy that emphasizes the careful construction of extensive data bases in a special field rather than rapid shift from one 'hot topic' to the next, longer but less frequent articles, and a reluctance to test hypotheses for fear of being shot down. The barriers to women are such that what appears to be a flawed strategy of reaction actually represents a creative response to obstacles in their path. We have found that in science, these strategies are enacted because the interpersonal networks that promote learning, the practice of the craft, the knowledge transfer, and ultimately the psychological freedom to take the risks inherent in innovative and creative work, are different for men and women. What is paradoxical is that while women pursue the myth that scientific individualism and isolation spurs scientific breakthrough, it is in fact a fiction that undermines their advancements, even as men (and some successful women) operate within networks of collaborative learning that advance ideas most competitively (Powell, Koput and Smith-Doerr, 1996).

SCIENTIFIC HEROINES
Even as they overcame the obstacles in their path, the most successful female scientists were constricted by their gender. The careers of Marie Curie, Lise Meitner, Rosalind Franklin and Rachel Carson provide us with benchmarks of how much has been achieved during the past century and how far the distance to equality was in each of their experiences. Indeed, the entry of women into scientific careers, as more than an anomaly, is a relatively recent phenomenon.

Just a century ago women were barred from seeking degrees and advanced training in the sciences in most universities in Europe. In their youth, during the late nineteenth century, Marie Curie and Lise Meitner received some of their training in so-called 'flying universities' through courses offered in the living rooms of homes by sympathetic male academics (Quinn, 1995). Other, less sympathetic, men believing that women's nature fitted them mainly for family and home, accepted female candidates only under exceptional circumstances, and still others, not at all.

When Lise Meitner emigrated to Germany from Austria to pursue a scientific career, she received financial support from her family that made it possible for her to pursue advanced studies. To Max Planck, the doyen of German physics in the late nineteenth century, Lise Meitner appeared to be one of those exceptional women and he allowed her into his advanced courses and, most importantly, his laboratory, a training experience that an improvised university could not provide (Sime, 1996).

During the nineteenth century women could attend German universities only as unmatriculated auditors. Baden was the first German state to open its universities to women in 1900. Prussia, where Lise Meitner aspired to follow her vocation for physics in Berlin, followed in 1908 and was by no means the last. Perhaps ironically, in the eighteenth century many laboratories, especially in chemistry, had been in kitchens in the home and thus more accessible to women's participation (Abir-Am and Outram, 1986).

The professionalization of the sciences and their incorporation into the universities during the nineteenth century placed the increasingly technologically sophisticated experimental sciences beyond the reach of most interested women. It was not until the 1970s that female access to the laboratory bench again reached the level that it had attained in the eighteenth century, a less institutionalized era in the sciences when upper-class women, at least, had open access to scientific work through their family and social connections (Gabor, 1995). Although women gained formal access to university-level scientific education in the late nineteenth century, informal barriers have persisted into the twenty-first century.

Such barriers are not so obvious as the rule that, even when she attained a research position, restricted Lise Meitner's presence at the Chemistry Institute in Berlin to a makeshift basement laboratory. Despite exclusion from the other laboratories and meeting places of her erstwhile colleagues, Meitner informally guided the investigations of male peers such as Otto Hahn through the force of her theoretical insight, combined with careful experimentation. Hitler's persecution finally drove her from her laboratory at virtually the last moment that a person of Jewish background could openly escape from Nazi Germany. Nevertheless, through clandestine contacts, she continued to advise her former colleagues on their research program. Always careful to soothe the male ego, Meitner negotiated a precarious path in German science, contributing at the highest level but receiving recognition at a somewhat lower level than her accomplishments warranted.

Meitner remained an outsider all her life, perhaps most poignantly during her years in Sweden, which provided a haven from Nazi persecution. Although she had a post at a research institute, she lacked access to support staff and research resources. Excluded from the Nobel Prize for the work she did with Hahn, Meitner received fuller recognition only late in life in the form of an Institute named jointly for her and Hahn, several individual scientific awards and a street named after her in Berlin. Nevertheless, she has perhaps only received full recompense from Ruth Sime, her excellent biographer (1996).

Despite the difficulties she encountered, Meitner was the key person in a leading German research center for much of her work life. Nazi persecution, and the war that marginalized Meitner, ironically brought another female scientist to the forefront. Until very late in her professional life, Maria Goeppert Mayer (later a Nobel prizewinner) pursued an outsider career even more on the margins of U.S. academia than Meitner's place in the German research system. Maria Goeppert grew up in an academic family in Göttingen and, when she showed an aptitude for physics, had access to leading scientific figures in the community such as Max Born who became her mentor. Nevertheless, when she married Joe Mayer, an American chemist, and moved to the United States in the early 1930s, her Ph.D. and advanced knowledge of theoretical physics only landed her an unpaid position in the physics department at her husband's university.

With his support and encouragement she was able to pursue a research career at the margins of Johns Hopkins University and then at the universities of Columbia and Chicago (Gabor, 1995). The war-time emergency that drew many women into the workforce also opened up a place for Goeppert Mayer in the Manhattan project, where her previous research meshed with the needs of the crash-program to develop the atom bomb. Until 1959, on the eve of receiving the Nobel Prize, when she left the University of Chicago with her husband to move to the University of California at San Diego, she held no full-time, fully remunerated academic position. She wanted nothing more than to be 'one of the boys,' fully accepted in scientific conversation.

To a great extent she achieved that goal. In discussions in the early 1950s with Enrico Fermi, the Italian physicist then at the University of Chicago, he encouraged her to formulate her ideas and set forth a claim to scientific recognition for her elucidation of the structure of the nucleus. Although she was granted a full academic position only late in life, Mayer can be seen as the prototypical traditional woman scientist, devoted to her work to the virtual exclusion of all other aspects of life. Only through far superior work could she be recognized as an equal.

Mayer's later career coincided with the beginning of the opening up of academic science to women's participation, often through pressures from the Equal Employment Opportunities process. Despite formal, tenured positions achieved by a growing minority of women, the way the world of academic science works still marginalizes women. Nepotism rules that prohibited universities from hiring husbands and wives were only the most overt of the many social and cultural restrictions on women's full participation in academic science. Nepotism rules are gone but reminders that science is a man's world persist even as women strive to make it their own.

In the early post-war era, when a London college's common rooms were still sex-segregated, men could take advantage of scientific women and get away with it by disparaging their femininity. This is how James Watson treated Rosalind Franklin in his autobiographical account, *The Double Helix*. Franklin concentrated on developing a data base of X-ray crystallography photographs to elucidate the structure of DNA but was reluctant to specify a structure until she could be confident of her results.

James Watson and his colleague Francis Crick were more willing to put forth speculative hypotheses but they needed access to her data to guide their model building efforts. Watson attempted to wheedle out the necessary information from her without offering collaboration and joint publication, the overt coin of the scientific realm. Rosalind Franklin, the co-discoverer of the chemical composition of DNA, relatively unacknowledged by her male peers and unavoidably passed over by the Nobel Prize committee, owing to her untimely death, had to wait for recognition from her biographer, Ann Sayre.

Rachel Carson, the biologist and author of *Silent Spring*, was widely recognized during her lifetime. However, her fame did not derive from research findings, in the traditional sense, but rather from analytical and literary accomplishments. Carson drew together and synthesized a broad body of evidence on the deleterious effects of chemical production processes and their effluents on the natural environment and human health. Indeed, Carson's own research career was stunted by the social environment of advanced academic science that made it difficult for a woman to find a Ph.D. advisor and be taken seriously as a scholar.

Despite her mother's unstinting encouragement and the availability of a female academic scientist (who herself experienced great difficulties in her research career) as a role model during her undergraduate years, Carson was precluded from a conventional research career by the obstacles she encountered as a graduate student at Johns Hopkins University during the 1920s. Instead, as is still the case for many women who wish to pursue scientific careers, she found a job at the outskirts of conventional science, in her case in a government bureau as a writer of pamphlets on ecology and wildlife.

Collecting the data for her writing projects through field trips and personal observation as well as from sources among a wide variety of researchers, provided the basis for her evocative and precise depictions of *The Sea Around Us* and other ecological themes that combined metaphorical insight and scientific acuity (Lear, 1997). Perhaps ironically, Carson's career on the periphery of science has become an exemplar of a new type of scientific career that emphasizes the relationship between science and society, rather than the traditional pursuit of research in isolation from its uses (Tobias and Birer, 1998).

Science writing, research management, technology transfer and science policy analysis are becoming careers in their own right rather than offshoots of research career lines. As science becomes more important to the political and economic spheres, the career lines that embody these intersections become less exceptional and more important. If traditional practices hold, however, one indicator of the increasing acceptance of such occupational endeavors will be their being taken up by an increasing number of men as well as women. If traditional discriminatory practices persist, the removal of women as leaders, if not practitioners, of these occupations, is also likely to take place.

UNSUNG HEROINES AND INVISIBLE BATTLES

Female scientists often told us, in interviews, about the obstacles that women encounter as they pursue their scientific callings. Academic practices, presumed to be meritocratic and gender-free, often work against women's professional success. These effects are sometimes hidden behind a neutral or even positive facade erected on the publicized achievements of a few exceptional women, some of whom deny the existence of obstacles in their path. Other women are unaware that they have been singled out for negative treatment while still others are all too cognisant but are also wary of challenging unfair practices for fear of reprisal.

Sex-role stereotyping sometimes colors advisor–advisee relationships. There are hidden obstacles, such as the length of the tenure process or the expectation that faculty members should move between schools to broaden academic training, which become apparent when a family or relationship is considered. Overt processes of discrimination include the sexual separation of scientific labor, with men seen as more appropriate to pursue the theoretical aspects of disciplines (usually mathematical not experimental) and women as more congruent with the parts of the field related to practice, policy, and the humanities.

The 1997 Harvey award lecture at Rockefeller University unintentionally symbolized some of the continuing gender disparities in science. The Rockefeller ceremony was a typical scientific honorary event in many ways. On the podium, the award recipient in black tie, Leroy Hood, the distinguished molecular biologist and professor of computer studies at the University of Washington, foresaw the union of the biological and computer sciences, and set forth the scientific, technological, commercial and health benefits that would issue from this marriage of disciplines. Curiously, even though Rockefeller University has a number of female faculty and graduate students, and the biological sciences have for some decades attracted a steadily increasing number of women, Dr Hood's formally attired cohort of hosts were all men.

Invidious distinctions, such as differences in timing, even appear in seemingly positive experiences such as the receipt of rewards. When a woman receives a prestigious fellowship or award, too often it comes late in her work life when it does not provide the same career boost as it would have at an earlier stage.

Cultural traits that are helpful to the conduct of science as well as those that are discriminatory must be disentangled from their origins in order to create a gender-neutral scientific role and workplace. The sexual separation of labor, the association of certain occupational specialties with one gender or the other strongly persists in most societies.

Perhaps ironically, the gender associated with a particular field may reverse, suggesting that the association is hardly inevitable. For example, nursing, a male occupation well into the nineteenth century, had become a largely female field not long into the twentieth century. The profession also, along the way, acquired the presumption of 'natural' association with the traditional feminine trait of nurturance. Those males who continued to enter the profession disproportionately assumed high-level positions, reflecting the continuing association of traditional male characteristics with leadership (Etzkowitz, 1971).

ECONOMIC AND STRUCTURAL BARRIERS

The state of the economy also affects conditions of entry and retention of women in science. Barriers to entry in industry and academia fall most easily under conditions of expansion and prove more intractable under conditions of recession. In the United States, Finland, and Portugal, women gained an increased proportion of R&D (research and development) positions during the post-war expansion of the sciences (Ruivo, 1987) On the other hand, when the expansionary period ended in Finland in 1983, it became more difficult for women, relative to men, to obtain posts in academic science. During such periods of increased competition, 'informal discriminatory practices and attitudes...' take hold with renewed strength (Luukkonen-Gronow, 1987: 196).

The renewal of discriminatory practices under harsh economic conditions can best be avoided if enough women have attained decision-making positions in science and technology workplaces by the time the downturn occurs. Otherwise, a disproportionate number of women ' . . . will lose their positions . . . unless preventive measures are devised' (Ruivo, 1987:390). Even when they retain their positions, a disproportionate number of women are to be found on the lower rungs of the job ladder in many scientific and engineering organizations.

OVERCOMING RESISTANCE TO WOMEN IN SCIENCE

Despite often having to put up a brave front in order to gain acceptance from their male peers, successful women scientists as well as other female professionals are becoming more willing to acknowledge the greater burden that they carry as women, and to seek changes in career structures and work styles. In an era of financial stringency and increased research competitiveness, change is made more difficult by pressures to obtain grants and lengthen one's list of publications. On the other hand, the struggle for equality is eased somewhat by allies among younger male scientists seeking some of the same reforms, to allow a better balance between their personal and professional lives.

Women scientists and academics, individually and collectively, are taking a more aggressive approach to redressing the imbalances between male and female participation, especially at the upper reaches of academia. Several generations of alumni of Radcliffe College are engaged in an organized effort to get the administration of Harvard University to increase the extremely low numbers of women with higher-level academic appointments at the university, including in the sciences. They have established an escrow fund to encourage donors to put their gifts on hold until progress is made.

The technical advisor to this effort is Dr Lily Hornig, a physicist and long-term activist on behalf of women in science. The perpetuation of gender-linked work roles and the continuing low rate of participation of women in many scientific disciplines appears to contradict one of the accepted standards of science: the norm of 'universalism', or in other words, the principle that scientific careers are open to all who have talent. The norm of universalism, formulated by sociologist Robert K. Merton, is that the acceptance or rejection of claims should not be based upon 'the personal or social attributes of their protagonists' (1973 [1942]: 270). It suggests that although science has traditionally been a male-dominated profession, it is not inherently so.

Moreover, by excluding persons of talent, as Merton argues in his analysis of the scientific profession in Nazi Germany, science is diminished by a 'racialist purge' (Merton, 1973 [1938]: 255). Although not as immediately striking as the elimination of Jewish scientists from German universities in the 1930s, the long-term relative exclusion of women has had a similar hampering effect on the conduct of science.

An earlier body of research identified as fallacious the notion that advancing age inevitably inhibited high-quality scientific work (Merton and Zuckerman, 1976). Unwarranted presumptions that youth was associated with high scientific achievement had served to justify extreme work pressures in early career stages. These unduly heightened expectations for early achievement have had unintended consequences on women's participation in science, given their coincidence with child-bearing years. Nevertheless, the implications of this earlier research for the structure of scientific careers, and the leeway for possible restructuring, has yet to be taken fully into account. We view these issues as 'critical transitions,' a series of overt and covert points in the life course when individuals are either propelled forward to careers in science or deflected away.

THE CONFLICT BETWEEN THE PERSONAL AND THE PROFESSIONAL

During their early childhood years, boys and girls develop different gendered images of scientists and what they do. Despite some early negative perceptions, large numbers of girls express interest in science and many follow up this interest through coursework and extra-curricular activities, often with the encouragement of teachers and parents. When they enter U.S. universities young women are dis-proportionately removed from science and engineering majors by a harsh 'weed-out' system designed to test the mettle of young males, well socialized in the norms of competition. Nevertheless, some women, looking back, report a positive experience of being mentored as undergraduates.

Despite the increased entry of women into science, opposition to their full participation continues. Implicitly 'male' standards of behavior permeate scientific time and space, including a belief that a researcher is most productive when their time is devoted to investigation to the virtual exclusion of all other aspects of life.

Ironically, the personal qualities required for success in science may be changing. Sociability, a trait traditionally associated with women, has also been found to be conducive to success in science, especially as the individual researcher is supplanted by group research, and multiple-authored publications become the norm. Perhaps, in the future, female socialization will become a career advantage in the scientific and engineering professions.

At present, female social attributes are a disadvantage that is exacerbated by competitive norms. While scientific training is an arduous process for all, our research and that of others suggests that women who aspire to scientific careers face barriers that do not equally exist for men and that equal success results only from truly heroic efforts (Abir-Am and Outram, 1987).

A letter to the editor of *The New York Times* entitled 'Science is for Childless Women' (May 17, 1995) exemplifies the persisting dilemma of women in science. The writer, Stephanie Dimant, identified herself as ' . . . one of those women who "leaked out of the pipeline"'. She cited the difficulty of reconciling the hours required of a bench scientist with the demands of raising a family. In a fast-paced, high-pressured environment, traditional solutions such as withdrawing from research for several years to raise a family and returning later were 'so unrealistic as to be comical.'

In bench science, ' . . . no second prizes are awarded, and the economic situation demands unrelenting writing of grant applications and publication of results.' Diment could not think of anyone she knew who had taken the extended leave option and who later returned to the academic track. Female scientists who made the decision to combine an academic career with raising a family typically took only the briefest time off for having a baby and then spent their limited maternity leave ' . . . with an infant in one hand and a telephone connected to the lab in the other.' Nor will there be many protests: given the stringency of research funding and the paucity of academic jobs, women do not want to be labelled as 'lame ducks'.

Nevertheless, given the pressures on women, including those that force the lower-paid spouse (rarely a man) to assume primary responsibility for child care, 'It is not surprising that many eventually make a heart-wrenching decision to leave bench science to those who have no children or to those who are fortunate to have that acknowledged asset, a wife.' Despite these obstacles, some women with children attain the highest levels of scientific achievement and recognition.

However, the price of success appears to be significantly related to each woman's ability to adapt to the highly competitive milieu of science. Dr Shirley Tilghman, director of a 'large and wildly successful' laboratory at the Howard Hughes Institute of Princeton University, concluded, 'Maybe it's because I've been in science so long that competition just seems like life. Maybe I've just given up.' A competitive sports enthusiast as a child, Dr Tilghman, a Canadian citizen and a recently elected foreign associate of the U.S. National Academy of Sciences, was featured in an article in *The New York Times* 'Fighting and Studying the Battle of the Sexes With Mice and Men.'

The article discusses Dr Tilghman's research on genetic imprinting, her experience as a mother and her concerns about the future of women in science. She is described as having 'jury rigged the pieces of her life by being almost preternaturally organised and focused, as well as spiritedly fierce in her work', in contrast to many women who draw back when criticized. Although she raised her two children as a single parent for most of their childhood, the article did not detail the child-care arrangements that made this possible.

Her own female graduate students were highly skeptical of their ability to follow her example, fearing, like Ms Dimant, that they would be forced to choose between science and motherhood. The students told Tilghman, 'Don't tell me about your experience. Your experience has no bearing on me.' They feared the time pressures of a highly competitive research funding system as well as the accepted belief that constant presence in the laboratory is a prerequisite for scientific success. Is there a one-to-one relationship between time put in and results achieved? Dr Tilghman attempted to reassure her students that '[h]ow one does in science is really dependent on your creativity and originality, and not how many mini-preps you can do in a 24-hour period', but the students were not convinced.

Unsure that this assessment applied to them, the students believed that the grant environment, now more competitive than their mentor had faced as a young scientist, inevitably increased the time that had to be devoted to a scientific career. Although a competitive environment also affects men, increased time pressures have additional effects on women. Thus, even this notable success story of a woman's achievements at the highest levels illustrates the persisting dilemma of women in science. This dilemma has its roots in the earliest years of childhood, and our next chapter focuses on how gender socialization affects the entry of girls and boys into scientific careers.

3 Gender, sex and science

The strong effect of culturally defined gender roles persists in science and other traditionally male professions through the social meanings attached to gender. Rather than a fluid perspective of human attributes that can be held by members of either sex, behavioral characteristics are frequently presumed to be innate and immutably 'masculine' or 'feminine' in the same way as one's biology.

The thesis that science is masculine, with 'masculine' understood as a cultural rather than as a biological term, ties issues of women in science to broader questions of gender roles and how they are culturally defined and transmitted from birth (Ruskai, 1990; Hyde, 1994). As Howell (private communication) points out, 'Sex, which is concrete and universal, specifies no role whatsoever.' Rather, it is cultural prescriptions and proscriptions, delineating which behaviors are appropriate to one sex and not the other, that creates the 'psychological meaning' of what it is to be male and female.

Thus gender as a concept was created to understand 'the social quality of distinctions between the sexes . . . for the explicit purpose of creating a space in which socially mediated differences can be explored apart from biological differences' (Hare-Mustin and Maracek, 1988). However, the concepts of sex and gender become easily entwined and socialization becomes confused with biology. Taking this a step further, ' . . . it would be illogical to say that being male or female would, in itself, make someone a good or bad scientist. Yet this kind of statement is often made.' (Howell, private communication). Negative stereotypes persist. The images of the role of women in science may be slightly more positive, but they have not been radically reshaped.

As the role of women has shifted to meet both society's needs and their own, some mothers and fathers relate to their daughters differently than in the past. They convey possibilities and expectations that transcend traditional role designations. Many of the young women whom we have interviewed over the course of the past decade not only report that it was their father who encouraged them to attain the Ph.D. in a science discipline, but credit their male advisors for sensitivity to gender inequities and their strategic assistance in helping them move forward. Individuals who encourage an interest in science need not belong to a particular sex or be a member of the family. What is essential is either a broad, flexible and encompassing vision of gender that incorporates non-traditional occupations or, paradoxically, a definition of gender in which it is viewed as irrelevant to vocational choice. The following discussion illustrates how far we are from this goal.

GENDERED CHOICES

The gender roles that children internalize influence which sex will choose to do science as well as who will have the best chances for scientific success. Blatant discrimination may be a thing of the past, but culturally generated gender beliefs play a significant role in leading children toward or away from an interest in science. Perhaps the most effective covert barrier to women is the simplistic idea that science is men's work and that women cannot make good researchers. The erroneous view of biological sex and gender as one and the same has led to the association of the male with the scientific role in western culture: science, like the Church, has been viewed as a 'world without women' (Noble, 1992). In most of this book we explore the conditions faced by women already in science. In this chapter we discuss the forces that work to divert females away from scientific careers from the earliest years of childhood through adolescence.

Differences between boys and girls appear at an early age as part of the social creation of the 'self'. As classically formulated by philosopher George Herbert Mead (1934), the child learns to 'take the role of the other' in play and other social relations. The self is thus constituted through a reflexive interplay of mirroring events. In the words of classical sociologist Charles Horton Cooley, the self is a 'looking glass self.' Much of our fate thus depends upon what other people think of us and how we respond to them. Children are influenced by the appraisals of others and respond according to those appraisals. If the information received is restrictive, whether based on race, sex or any other variable, there will be loss of human potential. Too frequently experiences for females are constricted through a process in which gender differences are recast into gender stereotypes.

Messages from those close to the child are especially influential in initially shaping a self-concept. As the influence of the child's family decreases, peers, authority figures, and culture as interpreted through the media, perpetuate the transmission of ideals of masculinity and femininity (Ortmeyer, 1988). The curiosity of infants and young children creates for them energy and excitement as they interact and are drawn to novelty in their environment.

However, if experiences are foreclosed and the child's world becomes constrained by what is seen to be appropriate to that sex, the child not only tends to abandon socially unacceptable interests, but comes to fearfully avoid that which is unfamiliar (Schachtel, 1959). Such a self-limiting process is exemplified in the historically high proportion of girls who lose interest in how the natural world works. When socialization impedes the individual's fulfillment of his or her potentiality, society as well as the individual loses.

VERY YOUNG CHILDREN'S CONCEPT OF THE SCIENTIST
Given the forces that push girls and boys apart, is there an inevitable dichotomy between the female and the male; the female and the scientist? A sample of fifty-three children from Southeastern Montessori School, ages two to six, were interviewed to analyze the emergence of gender differences in the perception of the scientist in early childhood. The middle-class school composition promised the optimal probability for a child's acquaintance with scientists and/or representations of science. A table contained four photographs (from the covers of *Chemical and Engineering News*) of male and female

scientists. The interviewer asked each child to tell her about the pictures: 'Who do you think these people are?' 'What are they doing?' 'How do you feel about them?'

Preliminary data that we have collected suggests that sex-typing persists and appears to become more evident the older the child. For some boys, science was seen as an activity that males, but not females, should take seriously. A typical response was that of a four-year-old boy who said, ' . . . only boys should make science.' The strength of the male identification with technology was also indicated by a boy who referred to a picture of a woman at a computer as 'he'. Yet in several instances rigid classifications by sex appeared to be less fixed as some of the children were able to identify both sexes with the role of the scientist. A four-year-old boy recognized a female scientist in the pictures and described her work thus, 'That one looks like a doctor.' 'She's working.' 'Something in a science. She's looking. Doing gravity. Making things fly. Someone who makes things we never saw before. With machines.'

In addition to discerning gender differences among very young children in their image of the scientist, the objective of this investigation was to identify discrepancies between their perception of the role of scientist and the child's view of themself. Boys were more likely to see themselves like the scientist, engaged in 'serious' behavior.

Boys, in general, were more negative in their views of women scientists than girls. Moreover, the older boys in the sample (ages five to six) were increasingly less likely to see girls as possible future scientists. One said, 'My sister Amanda wouldn't like to do this; she's really into Barbie dolls.' When, as part of the survey, the children were asked to draw scientists, more of the girls who drew women scientists did so in their second drawings. This suggests that even where the image of a woman scientist is held, it may be considered 'not quite right' and be presented only after the first, more acceptable picture of the male scientist has been recorded.

These perceptions and self-concepts illustrate the notion of the construction of gender schema (Bem, 1983), a highly selective process comprising 'a sprawling network of associations' in which information is taken in and organized according to the sex-differentiated values of the culture. A schema functions as a cognitive structure, serving to anticipate and make sense of new information coming in based on pre-existing perceptions. For instance, boys and girls at age two had 'concepts' of persons by occupation only if they had previous exposure to people holding the particular role. None of the two-year-olds were able to identify 'scientists', but those youngsters whose parents were expecting babies understood the notion of 'doctor' and thus applied this familiar concept based on the white lab coat. In other words, the young child's perceptions are not always dictated by a concrete situation. In this same way, the child learns to link attributes with their own sex or the sex of others. The perception of gender does not depend on the actual situation, but rather on organizing information that makes sense of novelty.

Two-year-olds without expected siblings, and therefore less exposure to doctor visits, were consistent with Piagetian developmental theory. When a child said, 'I don't know this person', it indicated that there were no available mental concepts whatsoever which reflected the attributes in the photograph. Therefore the child could make no interpretation of the photograph.

However, children above the age of three could identify scientific and medical occupational roles and had begun to link occupations with sex based on their knowledge of their family and the outside world. A three-year-old girl said, 'That's a man doctor,' while a three-year-old boy identified another picture as 'a big girl doctor with a cigarette.' A six-year-old boy said, 'My daddy's a builder; my mom's a scientist, but she's a student.' On the other hand, although a three-year-old girl recognized and characterized the activity, she did not attribute the activity to the woman in the photographs, only the man: 'Someone is working hard. He's a scientist because he is doing science stuff.' By age three to six, not only were most children in the sample familiar enough with the concept of the scientist to correctly identify the pictures, but they had begun to generalize sentiments and meanings across situations, demonstrating how at this early age information is encoded and organized according to cultural definitions of what is masculine and feminine (Bem, 1983).

In responding to the pictures, girls tended to see doctors while boys saw scientists. Showing the power of popular culture on gender images, one six-year-old male associated the images in the pictures taken from *Chemical and Engineering News* with similar images seen on television: 'They're called scientist[s]. I know that because of a cartoon that does [the] same thing he does.' Possibly boys are more frequently exposed to science in the media, games and toys, whereas girls apprehend the physician's role through their own increasing encounters with women doctors. When queried, girls more often identified the scientists as doctors, possibly because more obstetricians and pediatricians are now frequently women, chosen by other women rather than male doctors.

Young girls then not only learn about doctors in relation to female reproduction, but may experience this specialty as gender-neutral or women-friendly. Certainly it supports Bem's suggestion that gender schematic processing is dependent upon the social context and is a 'learned phenomenon and, hence, neither inevitable or unmodifiable.' Indeed, some boys draw similar androgynous conclusions especially when their mother holds a non-traditional occupational role. Thus, a three-year-old boy said, 'My mom's a doctor. [The person in the picture is] a doctor because he has a thing on his coat.'

Despite the persistence of sex-typing, there were indicators of change. For instance, a four-year-old girl demonstrated a working sense of the disciplinary order and division of labor in science. She said, 'Doctors fix people. A scientist checks things. I only want to be a veterinarian', and a four-year-old boy used clothing, not sex, as the identifying marker: 'Scientists! The clothes look like scientists.' Lastly, a six-year-old boy could identify researchers with a purely operational definition of the scientific method, notably gender-free, 'These are scientists. They're working really hard with experimenting to see if something does it or not. That's figuring out what do.'

Adults who provide a neutralizing message serve to counteract stereotypical notions of gender pervasive in the larger society. For example, a six-year-old female at Southeastern Montessori School told her interviewer 'I would like to do science. My mom gives me kits sometimes. A scientist is somebody who creates things. I like computers.' Although science was still linked to notions of gender by a number of the youngsters, some shift in traditional gendered associations of the scientific role can be inferred from the responses of others.

GENDER DIFFERENCES IN THE EARLY YEARS

The notion of clear-cut sex differences in the way newborns behave has not been borne out. In repeated studies girls could not be distinguished from boys without seeing physical differences. In those few studies where subtle behavioral differences were discerned, they were deemed moderate at best and were observed to diminish, disappear or be heightened based on the social context. The majority of studies have found that for parents, the sex of a newborn is a 'central organizer, a potent description of who the newborn baby is' (Tronick and Adamson, 1980). From infancy, boys and girls receive divergent messages from adults. Both Block (1984) and Hoffman (1977), in their studies of child-rearing practices, found parents encouraged their sons to actively explore the physical world, emphasized achievement, competition, and self-reliance, and felt it was important they try new things. In contrast, daughters were expected to be 'kind, unselfish, attractive, loving and well mannered [and grew] up in a more structured and directive world than males' (Block, 1984).

With few exceptions, these studies reflect how adult reactions to babies based on presumed sex perpetuate cultural beliefs about masculinity and femininity. Beginning in infancy, adults speak to and touch girls and boys differently. Sex-typed toys are offered by both parents with 'boy toys' providing more active physical manipulation and feedback from the physical world. In general, boys are given more freedom and less supervision, while girls are interrupted more frequently by their parents, particularly fathers.

Researchers alternated between dressing the same baby in pink and calling her Beth or dressing her in blue and calling her Adam. The adults who played with 'her' or 'him' noted that Beth was 'feminine and sweet' and Adam was 'sturdy and vigorous' (Will, Self and Datan, 1976). In a similar study, subjects were shown a videotape of a nine-month-old baby playing with a jack-in-the-box. When the baby was said to be a girl, her response was called fear; when a boy, anger (Condry and Condry, 1976) When a researcher told subjects that her new baby was a girl they responded with coos about 'how sweet she is'. When others were told that the baby was a boy, they said he was 'big and strapping'. In yet another study, parents were given a fourteen-month-old with whom to play. When designated a boy, 'he' was encouraged in active play with typically masculine toys. As a girl, 'she' received more nurturance and cuddling. In each instance, parental attitudes were projected onto the baby, depending upon the sex label, once again demonstrating the thesis of sociologist W.I. Thomas that 'a social situation is real if it is real in its consequences.'

These unconscious parental behaviors create an underlying 'gender awareness' during early childhood in which the world becomes categorized into mutually exclusive classifications (Condry, 1984). Gender roles come to be perceived as 'all or nothing' categories leading to prescriptions that scientists are men and secretaries are women. A female child may therefore believe that she cannot be a scientist even if she would like to because she is of the wrong sex (Kohlberg, 1966). This sex-typing process frequently continues with all authority relationships as the child's social world expands. However, the broadening of experience can sometimes provide the possibility for new influences which serve to enhance a child's self-concept when early familial experiences may have been rigid and stifling.

STEREOTYPING OF SCIENCE IN THE
PRIMARY SCHOOL YEARS
Among the many forces working against women's participation in science is the masculine image of the scientific role that frequently has

taken hold by primary school. This is often followed by neglect and/or discouragement of girls from doing mathematics in secondary school in concert with parental (and particularly maternal) perceptions of mathematics as difficult and non-essential for their daughters (Eccles and Jacobs, 1986). Moreover, conformity to stereotypes is frequently, but subtly, encouraged by educators and other authority figures. By puberty, these cumulative cultural messages are reinforced by the powerful need for peer approval and acceptance.

Unless the child is exposed to a wider range of possibilities, they may come to see gender-determined life choices as mutually exclusive and exhaustive. It is notable that both brighter girls and boys have come from family environments which were responsive to all aspects of the child's personality. Optimal performance was reported when children of each sex were encouraged to take the role of the other. Girls were free to actively explore and were 'encouraged to fend for themselves' while boys had received ongoing 'maternal warmth and protection' (Maccoby, 1966). However, once in school, many female children who have had a wholesome beginning in which they were relatively free to explore all aspects of themselves may now experience an erosion of the self based on stereotyped demands of teachers.

Enlightened parents may feel helpless in counteracting ubiquitous sexual stereotyping once their daughter enters school and other social situations outside the home. The recent dismay of dynamic classroom teachers who had allowed their interactions with students to be videotaped underscores the unconscious pull of stereotypical sex-typing behaviors even by educators who thought they were self-aware. As had been found in earlier studies, the teachers observed themselves calling on boys more frequently, providing extended conversation, information and help. On the other hand, because girls were less vocal and more cooperative, teachers were less likely to notice them or reward their talents, appreciating the girls' compliance in large classroom settings which they had to control.

A longitudinal study, over a 25-year period, found that girls were eight times less likely to call out comments, but when they did were reminded to raise their hands. In contrast, teachers responded to the typically rowdier and more assertive behavior of boys. Thus highly intelligent young girls often give up their own assertiveness and risk-taking behavior in order to earn their teacher's acceptance, fulfilling the social virtues of selflessness and cooperation (Sadker and Sadker, 1994).

By treating boys and girls differently, teachers encouraged 'the exploratory, autonomous, independent mathematical skills associated with males . . .' and discouraged them in females (Birns, 1976). Similarly, teachers gave extra attention to boys who chose to play at more complex tasks, but did not reward girls for the same behavior (Fagot, 1978). In addition, girls received few rewards for highly active behavior, whereas boys gained the attention of their teachers and also the admiration of their peers. While compliance may provide some rewards, it does so at a cost to development. Paradoxically, behaving like the boys can bring with it severe penalties, including adult reprimand and peer ostracism.

Significantly, teacher attention favors those attributes considered as male. Boys frequently succeed in gaining attention by using negative and inappropriate behavior, while less aggressive but more appropriate expressive bids by girls are often ignored (Block, 1984). Thus many teachers unconsciously reward compliance and cooperation from girls, while encouraging or condoning a highly competitive style of interacting for boys. At its most virulent, competitive 'putting others down' can become a pervasive part of personal interactions within the classroom. Not surprisingly, it is these very kinds of behavior on the part of adult male peers that have been identified by female Ph.D. candidates as disturbing and alienating.

Since boys have generally been previously exposed to manipulative toys, such as construction sets and models, they enter science classes with more confidence based on these earlier experiences. Moreover, the teacher's interactions with students influence children's perception of their own ability to do science. Not only are interactions more frequent with boys, but science experiments are often segregated into all-girl and all-boy groups with the boys receiving more attention, or if in integrated groups, the girls often watch the boys do the experiments (Wellesley College Center, 1992).

A learning environment that emphasizes experimentation, self-motivated exploration and inquiry is often an unfamiliar experience for girls, given 'the more structured, supervised, prescribed and proscribed world of girlhood as compared to boyhood' (Block, 1984). Thus rather than having a lack of interest in science, girls may tend to avoid the lack of structure in science laboratories where their anxiety would be higher than that of boys. In contrast to this kind of early experience, our studies indicate that women Ph.D. candidates frequently identify a particular high school science teacher who was attuned and responsive to their interest and inherent competencies, ultimately assisting in the development of independent exploration.

On a deeper level, these socializing processes may also account for later differences in the cognitive strategies used by girls and boys. In their more highly structured play and learning environments, girls use 'assimilative strategies' for adapting (that is, they tend to fit new information or experiences into their pre-existing cognitive ways of understanding) and are discouraged from engaging in more anxiety-provoking innovative efforts (Block, 1984). In contrast, where boys have been encouraged to explore 'a less predictable world . . . success in inventive ad hoc solutions would be expected to benefit boys' self-confidence' and the freedom to take risks. Along these lines, women graduate students in our studies, comparing themselves to male peers, reported feeling 'less able to take risks.' However, their prior educational successes and continued positive movement through the pipeline appear to reflect the importance of women's programs, as well as past and current mentors who created and continue to provide learning and social experiences free of gender-laden constraints.

Thus, gender differences become significant when masculine and feminine are defined in terms of narrow cultural norms, some of which are peculiar to American society. These norms set up a supposed contradiction between the traditional notion of 'scientific values' and what are considered feminine characteristics. Once they have identified themselves as males or females, girls and boys then want to adopt the behaviors consistent with their newly discovered status. This process of socialization results in children forming a conception of maleness and femaleness, revolving about such highly visible traits as hair style, dress, and occupation. They then use these gender images to organize their behavior and to cultivate attitudes and actions associated with being a boy or a girl (Kohlberg, 1966)

Although a young girl may have attended the same classes as male peers, by the time a young woman enters high school, she will not have had the same educational experience (Eccles and Jacobs, 1986). The cultural message that has predominated has been that active exploration, the capacity to be competitive, and the opportunity to handle machinery, play with chemistry sets and operate a computer are exclusively male activities. Not only are boys pictured prominently on the packaging of most science and building toys, but even the box of a chemistry set shows a girl looking on while a boy conducts the experiment. An eighth-grader described a dream in which she saw herself working for a scientist who did the experiments while she was left to write the paper. The youngster ended with, 'That's the way it is, right? . . . That's how we'll end up, the girls.' It is not that most girls will have been directly told that they 'can't' do what boys can. Indeed, most will be encouraged to 'fulfill their potential.' Nevertheless, in various ways many receive veiled messages of discouragement and denigration (Orenstein, 1994).

DISCOURAGEMENT OF GIRLS' INTEREST IN SCIENCE
DURING ADOLESCENCE

One of the primary tasks of adolescence is the further consolidation of identity. Peers replace adults in importance, and social acceptance has primacy. The cumulative subtle and covert messages regarding expectations and perceptions of females eventually influence the sense of one's place in the world, feelings of self-worth, and possibilities for the future. At a time when peer acceptance is crucial, conformity to stereotypical social roles is heightened. The anticipation of rejection by male and female peers in competitive activities may make such activities too threatening for most young women to undertake, particularly if such 'masculine' behavior contradicts the attitudes of the popular culture and the family (Block, 1984; Eccles and Jacobs, 1986). Therefore, adolescence is a logical time for the hidden meanings of gender roles to solidify further and become enacted in school performance and life choices unless the social milieu strongly encourages an inclusive ethos free of assumptions about gender.

By the time young women enter college many are not sufficiently qualified to major in those hard science areas that require a strong mathematical background, having avoided advanced classes in high school. Based on the research available and the fact that girls demonstrate equal mathematical capacity before adolescence, we can regard this 'giving up' of a potential skill as part of the legacy of the days when women's attempts at mastering mathematics met with indifference or overt rejection.

The controversy over differences between males' and females' mathematical skills, in concert with the issue of 'math anxiety', is of particular importance in illustrating how gender-appropriate roles affect later competencies. Nevertheless, an analysis of the results from numerous studies found no differences between males and females of any age in ability to understand mathematical concepts (Hyde, 1994). Gender differences in science achievement do not appear until the eighth grade; thereafter, strong gender differences in career orientation emerge, with half as many girls as boys showing interest in mathematics and science careers (Catsambis, 1994).

Thus, females appear to have the same aptitude for mathematics as males, but begin to lose interest and take only the minimum requirement in high school. Based on class grades, girls and boys are similar in mathematical and scientific ability until about tenth grade when girls decline to take elective mathematics courses. It is then that sex differences in problem-solving abilities begin to emerge. The question is not that of inherent ability, but one of why girls drop out of mathematics courses in high school and college (Hyde, 1994).

A 2-year longitudinal study of seventh- through ninth-graders, and their mathematics teachers and parents, argued that sex differences in attitudes toward mathematics, as well as achievement, are due to 'math anxiety' (Eccles and Jacobs, 1986). The level of anxiety correlates with the gender-stereotyped beliefs of parents (and particularly mothers) and the values placed on mathematics by the family. Students' attitudes and plans to continue taking mathematics courses were substantially influenced by parental perception that mathematics was difficult and of little value for their daughters. Thus rather than grades and performance having direct bearing on girls' self-confidence in mathematics, beliefs about their competence and desire to pursue interests and goals appeared to be strongly influenced by the parent's response to their daughter's grades. In short, prior performance, even when stellar, was secondary to the response of significant others. While there is a correlation between teachers' attitudes and the student's beliefs, the impact of teachers was not as strong as the influence of the parents.

There is conflicting evidence over whether the support of the school or the family is more significant to the minority of young women who do express an interest in science. Alice Rossi (1965) noted that a young girl with high intelligence and scientific interests must come from a very special family situation and must be a far rarer person than the young boy of high intelligence and scientific interests. On the other hand, if she reaches adolescence with the same intellectual inclination, it is often despite her early family or social experiences rather than because of them. This may reflect why women, when questioned in college about the background of their science interests, frequently point to particularly important teachers they had, often as early as the third or fourth grade, who provided them with new channels of communication and new expectations. In contrast, graduate students in our studies frequently cited both parents, and most often the father, as highly encouraging. Fathers have been found to be particularly supportive of their daughters' mathematical abilities (Eccles and Jacobs, 1986).

The gap in gender differences in standardized testing and achievement has narrowed since the early 1980s (Berryman, 1983; Orenstein, 1994; Hyde 1994). Nevertheless, girls' historically lower mathematics scores are typically attributed to biological characteristics, even a so-called 'math gene.' Yet how can biological differences explain the 50% reduction in this score gap between boys and girls by 1994 and current near-parity? Girls' supposedly lesser 'spatial abilities' have also improved with increasing exposure to spatial tasks. If mathematical skills were biologically determined they would presumably be impervious to such rapid and dramatic shifts.

Another seeming anomaly is the fact that non-Caucasian girls outperform boys in the highest-level mathematics classes in Hawaii (Orenstein, 1994). Since most people construct their perceptions of the world largely in accordance with cultural prescriptions that they take for granted, ethnic variation within the larger structure speaks to the influence of subsets, including family attitudes. During the past few decades there has been increased awareness of the way girls are treated in mathematics and science classes. Indeed, given the coincidence in the timing of the change in testing outcomes, even a modest shift in social attitudes might well be an indirect cause of improved awareness in how teachers relate to girls.

Nevertheless, it has been argued, perhaps most prominently in an article in the journal *Science*, that innate, biological male superiority was the best explanation for sex differences in standardized testing, based on the premise that girls and boys received identical training (Benbow and Stanley, 1980). In contrast to the fathers, mothers' confidence in their daughters' mathematical aptitude declined further in response to the *Science* article. Given the prestige of the journal and its prominence within the scientific community, the influence of the article in adding to pre-existing gender bias in the sciences could be high, but is still unknown.

Benbow and Stanley's paper certainly enabled advocates of 'meritocracy' to draw the conclusion that females are, in fact, 'incompetent' at science and competent at other things. Along these lines, our own findings indicate that any inference of 'difference', including variation in socialization, opens the floodgates for negative interpretation of what it means to be female. For instance, graduate women's programs, created to mitigate social isolation by building networks, are frequently interpreted as indicating that women have special needs. The possibility of different biological influences does not have to imply that behavior is 'predetermined'. Instead, biological propensities may be 'manifested in behavior in diverse and complex ways, as organisms are shaped by . . . the environment in which they must function' (Block, 1984).

Our concern is how socialization, based on stereotyped sexual division, appears to restrict the possibilities for girls and women and has a destructive impact on the sense of self for both sexes. The consequences of female socialization have a concomitant deleterious effect for young males as well, in the demand that they 'maintain the image of the aggressive, detached, active male' (Gould, 1978).

Neurological differences between right and left side brain development, or distinctive identifications and attachment between mother and daughter, have provided insight into the more verbal, relational and nurturing characteristics of females (Chodorow, 1978; Miller, 1976; Gilligan and Brown, 1990). However, qualities of maleness and femaleness are not rigid and impermeable. Boys and men have rich capacities for empathy, nurturance and attunement in relationships just as girls and women have aggressive, active, and competitive capacities.

Considering gender along lines of difference or non-difference presents numerous paradoxes with which women in the scientific community currently struggle. As mentioned above, when differences are acknowledged, as in graduate women's programs, females are negatively construed as the same (in some way needy or deficient) and not viewed as individuals. Thus, by focusing on difference, this approach minimizes similarities between males and females while obfuscating institutional sexism. On the other hand, adherence to a no-difference model 'makes man the referent . . . women must aspire to be as good as men' (Hare-Mustin and Maracek, 1988). For women scientists in less hospitable milieus, the no-difference model creates another paradox in which women must be either 'better than' or 'just like' men in order to prove they are equal.

FORECLOSING WOMEN'S CHOICE TO DO SCIENCE

By adolescence, gender socialization has affected career plans. The achievement orientation of the scientist depends on competitive success. Yet for females, competitive success is often accompanied by great emotional costs based on family attitudes and their early experiences in the classroom.

In many ways, women are unable to choose to do science; society has already chosen who will do science through its construction of gender roles. There is considerable evidence of the relationship between the adolescents' notions of gender-appropriateness and recruitment to scientific careers (Eiduson and Beckman, 1973; NSF, 1988). The image of the scientist as eccentric, non-conformist, and lacking in emotional capacity suggests that the potential recruit must have certain types of personality characteristics and live a particular lifestyle.

If the caricature of the scientific personality and lifestyle does not mesh with the student's interests, beliefs, and values, she or he is unlikely to become committed to being a scientist. For women, the requirements of a college major or pre-college mathematical preparation are not the only factors when making a career choice (Barnett, 1978). Rather, women avoid majors in science and engineering, in part, because they are socially ascribed as 'men's jobs.' Other studies corroborate these findings (Gerson, 1985; Berryman, 1983).

An early study by anthropologist Margaret Mead and Rhoda Metraux (1957), conducted during the 1950s, identified a negative image of the scientist among high school students in the U.S. and found that girls, especially, viewed a scientific career as an inappropriate form of work for themselves. Girls rejected science as being concerned with things rather than people. Moreover, they viewed science as a highly demanding career that would take them away from their future husbands and children, an issue which continues to trouble and impede women scientists today.

More recently, in Norway, strong clashes were found between girls' values and priorities and their perception of what it means to be a scientist (Sjoberg, 1988). Indeed, a series of U.S. studies, from the 1950s through the 1980s, showed that both boys and girls identify the typical scientist as a man (LaFollette, 1988). On the other hand, a study of elementary and middle school children in Taiwan found that although older students were more influenced by stereotypical images representing scientists as men, female students were three times more likely than male students to draw female scientists (She, 1995).

A study of the image of the scientist among older primary school students in Ireland found that girls, but not boys, drew pictures of female scientists, suggesting that even if the boys did not see science as an appropriate career for women, some girls, at least, could envision the possibility (O Maoldomhnaigh and Hunt, 1988).

Given these disparate and possibly contradictory findings across cultures, what is not yet known is which aspects of gender remain fixed and which are more flexible and amenable to change as individuals mature, particularly as they pertain to the image of the scientist.

American girls' performance in mathematics and science is still negatively affected by traditional gender beliefs. In other countries, particularly in Asia, boys and girls perform equally on mathematics tests. David Dunn of the University of Texas at Dallas notes, 'We tend, both in our family lives and in grade school and high school, to counsel girls away from math and science.'

By the time young women attend high school and college, they are frequently viewed as inappropriate persons to become scientists and engineers. Girls are often given the impression that they will face ' . . . intolerable obstacles, conflicts and handicaps' (Moulton, 1972), an understanding that all too accurately reflects the traditional organization of science.

4   Selective access

INTRODUCTION

Social practices that work against women's participation in science are often embedded in a seemingly gender-neutral competitive selection system. In this chapter we discuss how the normal workings of the U.S. higher educational system push women out rather than recruiting them into science and engineering careers. We contrast the workings of the unofficial 'weed-out' system in undergraduate education at large universities with a 'reverse weed-out' system at small colleges that must recruit students to their science courses in order to maintain their majors.

The weed-out system

In large universities at the bachelor's or first degree level, women often encounter a 'weed-out' system of courses based upon a competitive model that is designed to eliminate unwanted numbers of prospective students. This system has even worse effects on women than it does on men. Its encoded meanings, obscure to young women whose education was grounded in a different system of values, produce feelings of rejection, discouragement, and lowered self-confidence (Seymour, 1995).

A fortunate few women, after surviving this perilous journey, are recruited into a smaller scale, supportive version of the graduate research apprenticeship model. These women had no difficulties academically as undergraduates, in fact they were usually at the top in their classes and worked closely with their professors who were often important researchers. This perhaps explains why virtually all of the students interviewed in the graduate school samples reported positive and successful experiences in undergraduate school. Once in graduate school, many women recall their college experience as having been a nurturing environment that typically provided them with a mentor (as advisor, professor, lab director, etc.) who encouraged them to aim for the Ph.D. Ironically, once in graduate school, women often encounter a second weed-out system, a harsher, more discouraging, version of the research model they experienced as undergraduates. Their self-confidence, so precariously acquired in college, is once again deflated.

Most women who choose to major in science at university have had a positive high school experience which was one of the factors that encouraged them to continue. Thus, at each level the system removes disproportionately large numbers of women from the science career pipeline while providing a positive experience to a much smaller number, most of whom are fated to have a discouraging experience at the next level of their training. It may be said that the system applies to men as well, but as we shall see, the same strictures affect women worse than men, given the cultural differences between most women and men.

The weed-out system sifts large intake classes for intrinsic interest, talent, and fortitude, while, at the same time, drastically reducing the classes to a size that departments can handle in the upper division regardless of variations in the caliber of particular student cohorts. 'Weed-out' is a long-established tradition in a number of academic disciplines, but it is dominant in all science, mathematics and engineering (SME) majors. It has a semi-legitimate, legendary status and is part of what gives SME majors their image of hardness. It is thus an important feature in students' informal prestige ranking systems, both for individuals and for majors, disciplines, or sub-specialties.

Weed-out systems are similar to the 'hazing' practices of military academies and fraternities. Although these practices seem archaic they persist because they serve important functions that are difficult to achieve by other means. 'Weed-out' strategies are perceived as a test for both ability and character and are the main mechanism by which SME disciplines seek to find the most able and interested students of all who enter their introductory SME classes. The system operates in its most stringent form in larger, less elite universities, and although it still exists in elite universities and small colleges, its impact is moderated by countervailing forces. These forces include, in the one case, the likely higher class background of the students, and in the other case, the ability of programs to accommodate a higher proportion of students at the upper level.

The core of college education is the course, a set of class meetings held two or three times a week during a semester of fifteen or sixteen weeks, punctuated and/or concluded by examinations testing students' knowledge. The official purpose of a course is to impart knowledge to students, traditionally by lecture or recitation, more recently through laboratory practice or class discussion. Based upon the oral transmission of knowledge and originating with the founding of universities in the medieval period, before the invention of the printing press, the course of lectures has been a quintessential element of the academic structure.

In addition to its educational purpose, the course has traditionally had a role of evaluation, as the examinations attached to it show. Some colleges have tried to separate education from evaluation by scheduling examinations after blocks of courses, for example in a 'junior examination'. For the most part the examination has remained a part of the course, also serving as a sorting mechanism to place students into different categories. The highest category traditionally has been the few students most worthy of personal attention from the master: those most likely to have the abilities and inclination to become masters themselves.

As universities became training institutions for distinct professions the selection mechanism took on other functions as well. If there was a surplus of students interested in a profession, excess numbers could be selected out by raising the standards and eliminating the unwanted students.

Selection mechanisms can also accomplish more covert purposes, even some that may not be acknowledged consciously by persons running the system. For example, one covert goal may be to eliminate persons who are not in the image of those already in the profession. Selection can take place on a seemingly meritocratic basis by organizing the process according to cultural criteria that fit and therefore select for members of one group but are incompatible with, and therefore deselect, members of the unwanted group. Thus, the normal operation of the academic system will insure that reproduction of the profession occurs in a way that selects for people with similar social, cultural and economic characteristics to those already in the profession. Those eliminated will have little grounds for protest since the selection has seemingly been made according to universalistic standards.

The weed-out process acts as a post hoc selection system which avoids conflict with the ideal of open entry to higher education. Believing in the democratic ethic of the American educational system (which includes the idea that most people should be able to go to college if they have the desire and entry qualifications), most students were uncomfortable with the idea of decreasing access to college. In effect, academic faculty members are performing the traditional gate-keeping role of all professional bodies, from medieval guilds onwards, by identifying students best fitted for the profession, according to its own standards.

There are no references to weed-out systems in official university literature, and, when questioned, deans and faculty may be evasive, or deny their existence. Nevertheless, on entering the university, students soon become aware of a weed-out system. A previously unaware female student said, 'When I went to the orientation with my mom, the dean actually sat there and said, "Don't be surprised if about three-fourths of the people sitting here don't make it, particularly not in four years."' A more knowledgeable male student commented, 'They do the usual speech: "Look to the right of you; look to the left of you. Forty percent of you won't be here next year." I think that's the standard speech at every university.'

Weed-out systems also become evident to students in the ways that curricula were constructed, classes organized and taught, and assessment and grading practices set up and managed. A female student described her experience: 'The teacher was relatively young: I think he had just finished graduate school and he was kind of cold and cynical, kind of like, "I know a lot of you are going to drop out, so you might as well do it now, so that the rest of us can get on with this thing."' Student estimates of attrition targets ranged from 30 percent to 75 percent, with a median of around 50 percent. They also had a visual impression of how the weed-out process was progressing by shifts in the seating patterns in their classes. A male student said, 'You can always tell. There's what we call the "T", the students in the front two rows and down the middle, they're the A students, and everybody else you're gonna lose.'

GENDER SOCIALIZATION AND UNDERGRADUATE SCIENCE EDUCATION

Socialization into gender roles affects the educational experience, especially when teaching styles are skewed in favor of one gender rather. Highly competitive in nature, introductory science and engineering courses tend to select out women. Although highly motivated and scientifically able, women are not as accustomed as men to the rigors of competition and thus are removed from the career pipeline. The system for intellectual and moral education of young men in the sciences and engineering contradicts female expectations. Young women, who worked hard in high school and used their teacher's praise and encouragement as the basis for their self-esteem, become disoriented in college. Lacking experience with the 'male' culture of science and engineering majors, most women do not know how to respond appropriately. Women quite realistically sense that its standards differ from their previous experience and that many men resent their presence.

The disproportionate exclusion of women from the upper levels of science and engineering education is an important latent function of the weed-out system of the first two years of engineering and science majors. That women and men respond to the scientific education system differently is exemplified by a female student's observation, 'Science is a wonderful example of how men just have their own little world – just men, and men's ways, and men's concerns, and men's thinking.' The system does not relate to the way that women are taught to learn, nor to the models of adult womanhood that their socialization encourages them to emulate. Even well-prepared female first-year students enter basic classes feeling uncertain about whether they belong. Faculty members who teach 'weed-out' classes discourage the kind of personal contact and support that was an important part of high school learning. The loss of regular contact with high school teachers who encouraged them to believe in their ability to do science exposes the frailty of their self-confidence. (As we have seen, the relatively few women who avoid the debilitating effects of 'weed out' and advance to higher levels encounter a similar experience upon entering graduate school.)

The system tests for characteristics traditionally associated with 'maleness' in Anglo-Saxon societies and is based on motivational strategies, such as the idea of 'challenge', understood by young men reared in that tradition. Challenge is a central theme in many rites of passage into manhood: the boy is challenged to test his mettle against that of the established adult males who set hurdles for him to surmount before he is allowed to join them, initially as an apprentice, ultimately as an equal. The nature of the challenge is as much moral as it is intellectual, in that it is intended to test the ability of young men to tolerate stress, pain, or humiliation with fortitude and self-control. By a deliberate denial of nurturing, young males are forced to look inward for intrinsic sources of strength, and outward to bond with their brothers in adversity – their peer group.

Most faculty members in science and engineering departments treat young women the same as they treat young men. But this seeming equality actually differentiates against women in asking them to perform in ways that are contrary to their socialization. By 'challenging' everyone in the class to 'prove' themselves in the face of harsh teaching methods, rapid curriculum pace, and a rigid assessment system, academic staff send a meaningless message to the female minority. Not only is the metaphor of 'challenge' obscure to female students, so, too, are other elements in the traditional male educational process such as 'proving' yourself, a gender-defining activity for men that is risky and inappropriate for women. As one young woman said, 'I'm not going to waste any more of my time proving myself. I know who I am, and what I can do.' To be drawn into the male model is to court anxiety, insecurity, and confusion about the basis of one's sense of self.

Competing for grades is another aspect of the male testing process. It has ill effects on both women and men, though not necessarily for the same reasons. Competition is about 'winning', which is the most traditional way of placing individuals within male prestige and ranking systems. It is a central feature of all military, political, and economic activity, and is metaphorically represented in sports and games originally developed by men. As women increasingly involve themselves in these areas of activity, some women adopt the competitive imperative, and learn how to compete in male terms. Men are often not comfortable with this. It is their game, and there is no place in their prestige system for a woman who competes successfully with them.

The extent to which women adapt to the system depends upon the degree to which they have already accepted competition as a way of relating to others in high school, or in sports and games. Entry to first-year science, mathematics or engineering suddenly makes explicit, and then widens, what is actually a long-standing divergence in the socialization experiences of young men and women. The divergence in self-perceptions, attitudes, life and career goals, and customary ways of learning and of responding to problems, which has been built up along gender lines throughout childhood and adolescence, is suddenly brought into focus, and into practical significance.

The essential opposition between two categories embedded in the traditional gender-role system has consequences for all students and faculty members. It occurs when a relatively small number of inexperienced young women are encouraged (with little prior preparation in the cultural and personal dimensions of their undertaking) to venture into an institutionalized national (possibly international) teaching and learning system which has evolved over a long period as an approved way to induct young men into the adult fraternities of science, mathematics and engineering. Most young white men seem able to recognize, and respond to, the unwritten rules of the adult male social system. The rules are familiar because they are consistent with, and are an extension of, traditional male norms, established by parents and reinforced by male adults and peers throughout their formal education, sports, and social life. The same set of norms are to be found in the education and training systems used by many occupations and professions, including the military.

The ease with which young men adjust is variable; but the nature of the undertaking is, at least, familiar. Indeed, the ability of male students to recognize, and respond appropriately to, these male norms transcends national boundaries. For example, at one institution which regularly attracts students from Norway, a Norwegian woman in our sample commented on the ease with which her male Norwegian peers seemed to adjust to their engineering and science majors. She contrasted this with her own difficulties in developing a sense of belonging in her major – a difficulty which she shared with American women.

Many aspects of science and engineering majors force women into conflict with their gender socialization. The resolution of these conflicts is sometimes accomplished by leaving the major; sometimes by making personal adjustments to the dominant male social system. These adjustments tend to be psychologically uncomfortable, and some coping strategies provoke disapproval from other women, male peers, or both.

Most young women develop a sense of identity that is highly sensitive to extrinsic response. From very early childhood, throughout the years of formal education, girls are encouraged to perform to please others, and to base their feelings of confidence and self-worth on praise or other signs such of approval. The degree to which any woman depends on significant others for her sense of achievement varies according to her mixture of cultural influences. Nor is the tendency to perform for others restricted to women; depending on the circumstances of their upbringing and education young men may also exhibit this trait.

The ways in which women have learned to learn also raises the difficult issue of whether, and how, to change the traditional ways in which girls are socialized and educated. Even if we knew how to teach girls to be more independent in their learning style, is it desirable to change the collective identity of one gender group so they can more easily be fitted into educational settings which reflect the learning styles of the other gender; furthermore, some aspects of the learning environment in which women feel most comfortable – particularly learning through cooperation, interaction, and experience – encourage the development of skills and attitudes which have increasing value beyond academe, especially as the need to work collaboratively increases in science and business.

Part of the traditional socialization of women has been the development of a high degree of tolerance for behavior which is increasingly being redefined as 'abusive'. At a trivial level, this includes 'making excuses' for rude or insensitive male behavior in order to preserve the appearance of normal social or domestic relations. When talking about how they respond to rude peer behavior, female students made comments such as, 'It's best to just ignore them', 'Reacting just makes it worse', and 'They'll grow out of it.' Women who felt angry expressed it to each other, rather than directly to the men concerned.

Where the power differential is so much to their disadvantage, and there are no guidelines for responding to the situation, women fall back on learned ways of discounting abuses of male power. Assuming the traditional female role of 'peace maker' comes at the price of tolerating an abusive situation, and, in this case, of offering some rationale for that accommodation to the researcher who questions it.

A second possible explanation, which is not inconsistent with the first, derives from games theory. The outsider who wishes to become a player in a game which is already under way, with a group who know the rules, who are more skillful players, and to which he or she does not belong, has to accept admission tests – even if they seem silly or arbitrary. Although our women informants described the constant implicit demand of their male peers that they 'prove themselves' as foolish and irrelevant – as, indeed, for women, it is – they nevertheless were drawn into proving behavior.

They felt constantly forced to demonstrate their 'right' to belong, and part of their motivation to work hard, or harder than the men, was a vain attempt to force this concession. Those women who adjusted their presentation of self to a parody of male style can be seen as seeking to side-step the admission test by claiming group affinity. Paradoxically, while disputing that unpleasant male behavior bothers them enough to undermine their motivation, the female minority tacitly accepts the rules of the game imposed by the dominant group.

Women were also concerned that male acceptance of their academic worth would detract from their sense of who they were as women. The problems of belonging and identity are linked, because the qualities that women feel they must demonstrate in order to win recognition for their 'right to belong' (especially intelligence, assertiveness, and competitiveness) raise the anxiety that such recognition can only be won at the expense of 'femininity'.

Women are forced to make a cultural choice between being attractive and being smart. As one female student said, ' . . . maybe I was afraid to be too good at it . . . that if I showed how good I was, I would lose my femininity – that men wouldn't find me attractive. I think I've always been encouraged to mess up, then guys come and help you out [laughs] – even though I didn't really need the help. But they have to think that you do . . . Subconsciously, I really felt that if I succeeded, then they wouldn't see how attractive I was.'

To succeed in science and engineering, women are forced to follow the male model, but most women are reluctant to do so, with obvious implications for their willingness to remain in the profession. As one woman realized, 'It's set up that women have to be more male in engineering to get along. I notice that women in other majors don't seem like they have to change themselves like I did in order to fit in.' Women face a double-bind situation and can only win male acceptance, in academic terms, by losing it in personal terms. ' To make it in engineering, I had to learn to be more male . . . Eventually, you've learned to take more stuff – maybe are stronger than when you first came in. But it always bothered me that I had to change.' The extrinsic nature of traditional female identity is both defined and confirmed by men. Women can be set up to fail, unless they are helped to see how the existing male-dominant power structure can play upon their anxieties about their self-image, and are offered some strategies to protect themselves from it.

BEATING THE SYSTEM OR BEING BEATEN BY IT

A common theme that distinguishes the accounts of women and men in science and engineering majors is rupture with past educational and social experience. Notwithstanding the discriminatory pre-college experiences of some women, or the doubts generated by a generalized cultural discouragement from the pursuit of non-traditional disciplines, most women enter college in the U.S. at a peak of self-confidence, based on good high school performances, good scores in their Scholastic Aptitude Tests, and a great deal of encouragement and praise from teachers, family and friends. Soon after entry into college, women who felt intelligent, were confident in their abilities and prior performance level, and took their sense of identity for granted, began to feel isolated, insecure, intimidated; to question whether they 'belonged' in the sciences at all, and whether they were good enough to continue.

A female student whose confidence in her ability is highly dependent on the judgments of others finds it difficult to judge the adequacy of her performance. Receiving what are viewed as adequate or even good grades for their classes is not in itself sufficient to prevent what women commonly referred to as feeling 'intimidated' and 'discouraged'. Her self-confidence may be already shaken by her abrupt reduction in status. In high school, she was treated as special. Now, she is part of an unwelcome minority which is treated with a hostility that she cannot explain. Her new college teachers, to whom she looks for guidance, ignore her.

Part of the difficulty women experience in defining their performance as adequate to the task is their isolation. Without a support network of people with more experience, it is easy for each of them to assume that they alone are struggling. Even when their performance is adequate or good, women who have an under-developed sense of their abilities in mathematics or science have difficulty in knowing that they are 'doing okay' without the teachers' reassurance. Deprived of that exchange, certainty about self-in-science is lost until the relationship is re-constructed with another supportive teacher, or a more independent self-concept is developed.

For the first time in their lives, white women suddenly experience what it is like to be a minority, negatively viewed by the majority. A young woman said, 'It's intimidating to be in a class with ninety-seven men and just three women – at least, it used to be: I think I've finally gotten used to it.' From the outset, they are excluded from conversations and activities solely on the grounds of characteristics which they cannot hide, and over which they have no control. A young man commented, 'Women just can't break into those solid ranks of men. It may just be as simple as that. It's always been male, and they're gonna keep it that way.' Many men are well aware that they or their peers often exclude the women in their classes from their working or social groups solely because they are women.

Unfamiliar with this experience and lacking contact with senior women who understand the nature and source of their problems, first-year women find it difficult to make sense of their discomfort. As one young woman expressed her need for affiliation, 'I need to feel like there's someone there sharing it with me. I don't want to feel so alone . . . it gets you down . . . And, if you get down about something, it snowballs, because you've no one to talk to. That's when you get to the point of, "What am I doing here?"' Few had received any guidance about what to expect, and how to survive; they lacked a female folklore offering ready-made explanations or remedies for their difficulties; and most had little knowledge, or acceptance, of the analytical framework offered by feminist theory. In short, they were inexperienced eighteen-year-olds, who tended to blame themselves when people behaved disapprovingly towards them.

Since they are raised to work more for the approval of others than for intrinsic satisfactions and goals, many women fail to develop a clear personal view of what they want out of college before they arrive. This also explains why the openness of teachers to the personal approaches of their students is so central to women's definitions of the 'good' teacher. For many women entering college, engaging the teacher in a personal dialogue appears to be critical to the ease with which they can learn, and to their level of confidence in the adequacy of their performance. Failure to establish a personal relationship with faculty members represents a major loss to women, and, indeed, to all students whose high school teachers gave them considerable personal attention, and who fostered their potential.

To a much higher degree than is the case for young men, preserving the self-confidence which young women bring into college depends on periodic reinforcement by teachers. The prospect of four years of isolation and male hostility on the one hand, and the abrupt withdrawal of praise, encouragement, and reassurance by teaching staff on the other, depletes self-confidence. One young woman said:

> After the positive influences and positive reinforcements in high school, you feel on top of the world, and that you can do anything. Then you get into an entirely new system. I noticed a marked difference in my attitude. And I believe it was because of the fact I was a number and nothing else to anyone . . . I had no one to perform for – and probably many other women are so used to being performers for others, that you take that away and you're left with

> a void. And at the time, I didn't really know it was that. The classes I do best in are the ones where the professor cares about me, and it's always been that simple for me. I cannot separate my feelings for the professor from my performance.

Faculty members may, or may not, realize the critical role which they play in the persistence of women, both as a source of ongoing support, and at times of crisis. Many women offered 'fork-in-the-road' stories in which, having plummeted into depression, confusion, and uncertainty, they sought the counsel of faculty members about whether they should continue. They were prepared to accept their professor's assessment of their ability and performance, so long as this was conveyed in a manner that suggested he or she cared one way or another about their well-being. Describing a critical time when they felt unable to trust their own judgment about their ability to continue, seniors recounted the vital difference made to their decision to stay by expressions of support from faculty members whom they consulted.

The personal style of some college teachers, and their active, open encouragement of women in their classes, or in advisory sessions, made an enormous difference to the confidence with which women tackled their work, and, therefore, to their likelihood of success. If women survive, it is partly because someone noticed they had the talent and encouraged them in the first place. Even more important, they have received some support along the way. As one young woman summed it up, 'It's not any one characteristic in women that stands out as making them likely to succeed – like having lots of will-power or something. It's more that their talent has been supported. They've been helped to keep going, and not let the discouraging things get them down.'

Male undergraduates who meet the challenges presented to them in the early college years are assured of mentoring by the adult fraternity once the weed-out process is complete. Women who survive the undergraduate testing process do not automatically receive this reward. There is a seeming anomaly between our undergraduate and graduate data sets. The undergraduate study found that female survivors are often not accepted into the fraternity, except as tokens, or are not supported by it. The graduate study concluded that in women's colleges and small liberal arts colleges, and in some departments in larger universities, women do receive the support and mentoring that place them on the path to graduate school. An increasing number of women are entering graduate school in the sciences and engineering, drawn from the ever larger pool of qualified female BA recipients. However, with the exception of a very few scientific fields, a significantly larger proportion of men than women proceed from undergraduate to graduate school.

UNDERGRADUATE SCHOOLS THAT PROMOTE WOMEN'S INTEREST IN SCIENCE

Nevertheless, there are important differences among undergraduate schools in preparing women for graduate training. A female faculty member at a prestigious graduate school observed that women students from women's colleges appear to have greater self-confidence. From interviews with female Ph.D. candidates who emphasized their need for 'safety' in order to practice presenting papers and developing a professional self, it seems likely that women who have attended women's colleges have had the opportunity to take necessary risks in a secure environment while being supported by a committed faculty. Some female graduate students are curious about the behavioral differences they also perceive in classmates who have attended women's colleges: 'Confidence is the most important. It's what needs building. I've read that the women who go to women's colleges have much more confidence than those who have been competing with the males all along. I met a young woman in a class and she said she felt that it had made a big difference. You could just tell. She just had a different manner.'

Another female graduate student had developed an assertive style, yet maintained strong opinions on the needs of women. She had received solid mentoring before graduate school and felt she had benefited by having worked in a research laboratory directed by a woman, within a small university with four women Ph.D.s on faculty. Her capacity to look after herself was carefully developed by both male and female mentors before her entry into graduate school. Her laboratory advisor and professors primed and prepared her, teaching her strategies and the realities she would experience. Her subsequent ability to negotiate the graduate school system argues against hiding sexual discrimination from women before or upon arrival at graduate school.

She said, 'I absolutely was prepared. I worked for two years in a real research laboratory of a woman, one of four on faculty at the university I went to. Pretty much along the line she would say, "This is the kind of class you want to take if you want to go to graduate school." And when I started studying physical chemistry, my professor stated, "Now these are the kinds of things you are going to want to do when you go to graduate school." I actually had a professor take me aside and say, "Okay, now the rest of the world doesn't have four women on faculty." They tried to get me ready for the big world. They wanted to make sure the move wouldn't be a shock. So maybe they gave me the worst perspective and then said reality is somewhere in between. They always let me know, "We wouldn't be telling you to do it, if we didn't think you could do it." There was always the reality, but there was always the support. A lot of support.' Obviously, more women need to receive that kind of experience, and sensitive mentoring, in undergraduate school.

For contrast, we also conducted several focus group interviews with science students at a small state university college. The existence of a weed-out system was recognized in one course, organic chemistry, where students were aware that many who began would not finish. The dominant reported experience in virtually all courses was that professors were available to speak to students about their difficulties. A system of undergraduate teaching assistants was in place with regular meetings of students in a class held in small groups. The teaching assistants also encouraged students to form their own study groups. Given the small size of the science departments, the emphasis was on retaining students rather than trying to eliminate them. Indeed, with most advanced classes having fewer than ten students, the problem was too few students rather than too many. The rationale for a weed-out system was absent. Indeed, a 'reverse weed-out' system appeared to be in place in which students were strongly encouraged to complete their degrees.

CHANGING THE WEED-OUT SYSTEM

Too often, undergraduate teaching staff conflate the male role with the role of the scientist, to the predictable detriment of their female students. The more the faculty treat the demonstration of 'masculine' characteristics as an essential part of 'becoming a scientist', the more resistance women experience to their participation. This is the precise opposite of what many young women – and some young men – feel they require in order to give of their best, that is, teachers who care about them, advise on the adequacy of their work, praise or chide them, as appropriate, and give support through periods of difficulty.

Unable to evoke such responses from the largely male faculty (or from those female faculty members who have adopted the style of their male colleagues), women in science and engineering classes tend to feel they must be performing badly, and doubt that they should continue. Male peers advocate not taking faculty 'rejection' to heart. Many women have little experience of taking it any other way.

Young women tend to lose confidence in their ability to 'do science', regardless of how well they are actually doing, when they have insufficient independence in their learning styles, decision-making, and judgments about their own abilities, to survive the lack of motivational support and reassurance by faculty, or the refusal of male peers to acknowledge that they belong in science. Women who persist tend to have entered with sufficient independence to adjust quickly to the more impersonal teaching, have an intrinsic interest in the major and a strong sense of career direction, and develop attitudes and strategies (including alternative avenues of support), in order to neutralize the effects of male, peer hostility. However, the loss of many able women cannot be reduced without changing traditional faculty norms and practices (as well as those of some high school teachers and advisors).

The emergence of gender parity is also a spur to cultural change in engineering and science departments. In the life sciences, and some mathematics departments, female students report the atmosphere to be more comfortable, and the problems fewer. As one young woman reported, 'Well, in biology, it's fifty–fifty, so I just never felt that much of a difference.' Similarly, at the two research universities, and at the small liberal arts college, which were actively recruiting male and female students into science and engineering majors in more equal proportions, the discomforts caused by male peers and faculty were considerably less than they were in the same disciplines on the other four campuses studied.

While change is under way, first- and second-year women need programs to help them understand the source and typical nature of the discomforts and self-doubts they experience; strategies to deal with them; and support to off-set tendencies to self-criticism, sinking confidence, and emotional confusion. These difficulties are induced by normal educational experiences in science and engineering and are entirely predictable. Thus, programs for women in unremediated situations cannot be effective when they are set up on a one-on-one, crisis-based, 'women's advisor' system, or when they lack the public commitment of senior administrators and departmental chairs. (As we shall see, the same conclusion holds for the graduate level.) Successful programs draw on the involvement of senior women students, faculty women, and sympathetic male faculty members, in each major, and on a network of professional mentors.

In some departments, cross-cohort informational and support networks have been established by chapters of national societies such as the Society for Women in Engineering (SWE), and the Association for Women in Science (AWIS). Other strategies include: field-based residential options; pre-college orientation programs; mentoring systems (including pairing senior with more junior women); and augmentation of classes with all-women tutorials, seminars, and study groups. Some departmental and institution-wide programs which exemplify these strategies, are the residential program for women of color at Stanford, the WISE programs at Brown University, and, at the University of Washington, both the Women in Engineering Programs (WIEP), and the WIS and Freshman Interest Group programs for women in chemistry. The period over which such programs continue to be needed will be determined by the speed and profundity with which traditional attitudes and practices are addressed.

Changes are needed, not only in the transition from one phase to another but in the internal structure of each state of scientific career preparation. The culmination of higher education is attainment of the doctoral degree as a certification of the ability to advance knowledge in a field and a license to train others to become 'doctors'. In the next chapter we discuss the Ph.D. socialization process, how women are treated differently than men, and why.

5 Critical transitions in the graduate and post-graduate career path

Graduate education is not a smooth continuum, with a steady rate of 'leakage' from the pipeline, but rather a discontinuous, turbulent flow, with attrition rates rising at certain key junctures (NSF, 1994). We have identified several specific points in the career trajectory when people are propelled forward, pushed out, or dropped down to a lower level. We call these points 'critical transitions.' At the Ph.D. level these transitions are likely to include: (1) the qualifying examination, (2) finding a research advisor, (3) negotiating a dissertation topic, and (4) deciding what is sufficient work for the granting of the degree. Academic transition points sometimes coincide with events in the course of a life time that affect how decisions are made. Thus, for example, a student who is pregnant might have difficulty in finding an advisor, if decision makers view child-rearing and research as inherently incompatible.

The most crucial transition in the experimental sciences is the one from being a student in courses to becoming part of a research environment. A female graduate student described it as an apprenticeship: 'You learn the part of being a physicist through interaction with other physicists.' However, a belief permeates many departments, and is transmitted to incoming female students, that their admission is based on affirmative action rather than merit. Female students' self-confidence is eroded by the attitude of faculty and male peers that women are less competent than men. Rather than promoting interaction as fellow scientists, the attitude towards women can create feelings of incompetence and lack of success. As a female faculty member observed, 'This support [of the faculty] is key. If you don't have it, if you have people with the attitude that it's their job to fail certain people, then yes, people are going to drop out. Then they wonder, why don't we have more women here.' The ability to negotiate a transition point successfully often depends on access to informal sources of information which are often more readily available to males than females.

Transition processes are not uniform but are strongly affected by degree program organization and structure. The experiences of Ph.D. students vary widely, depending upon the practices of a discipline, university, department or advisor. For example, finding a research advisor in biology takes place through the custom of rotation among laboratories during the first year, which typically introduces the entering student to three professors and their research practices. Some critical transitions are highly structured, with clear benchmarks; others are more informal with loose or shifting criteria. A few Ph.D. programs have recently been reorganized to make transition points more flexible. For example, in some instances a series of written qualifying examinations have been replaced by a research paper and sets of course grades, opening up alternative paths to certify acquisition of sufficient knowledge to undertake a dissertation.

THE U.S. GRADUATE EDUCATION MODEL

In the mid-nineteenth century, when American scholars returned from Germany after earning their Ph.D. they attempted to replicate the advanced education system they had experienced abroad. Not surprisingly, a building project for a domestic institution was often an interpretation of a particular idiosyncratic professor's laboratory abroad. Although U.S. scientists founded research institutes according to the models they had learned in Europe, their efforts usually failed through lack of resources at home.

By the late nineteenth century, when the German-style hierarchical professorship failed to take hold as the model for organizing research and teaching in U.S. universities, the department was invented as a consortium of, more or less, equals (Oleson and Voss, 1979). The U.S. academic model was based upon a professorial status, with the ability to initiate research, granted early in the academic career.

The department became an association of relative equals, with each professor representing a different aspect of the discipline. The emerging U.S. academic system was much less centralized than the German model, typically built around a single professor. Research was developed relatively inexpensively by hiring students as research assistants instead of using Ph.D.s as in the European Institute model. In time the Ph.D. training process in the sciences was also transformed from an individualized research endeavor, which still persists in the humanities, to a group effort. In an apprenticeship format, an entering student typically takes off from the work of an advanced student and is, in part, supervised by the student whose work they are building upon (Etzkowitz, 1992). Under these collaborative conditions the dissertation is also transformed. Although still presented under an individual signature, the thesis increasingly looks less like a monograph on a single subject and more like a series of co-authored articles on discrete topics.

Some female graduate students assume that the old model of the lone investigator still holds. Often less integrated into their research group than men, they sometimes expect to have to produce a magnum opus for a dissertation. Some male faculty members, who are resistant to women, use this cultural lag against their female students by assigning ambitious projects in expectation of inducing failure. After the supportive social environment that many experienced in their undergraduate training, female students are often surprised at the resistance to their presence in graduate departments. Lacking access to informal sources of information that would allow them to make a smooth transition, women usually find the norms and rules of graduate school opaque and difficult to decipher, placing them at a severe disadvantage.

Contrary to gender stereotypes, female graduate students are often left to be the 'rugged individualists', having to fend for themselves, while male professors draw many of their male graduate students into a supportive, caring environment. Such coteries surrounding a faculty member, typically including students from many nations and cultures, constitute the basic social unit of U.S. doctoral education. The countries that Ph.D. candidates come from may be at odds in the outside world, but in the research group students, irrespective of their background, are expected to form bonds that will last a career, if not a life time. In the experimental sciences there is typically a common physical site in a laboratory. In theoretical fields the informal social ties that form the basis of the group often originate in a seminar.

THE ILLUSION OF MALE AUTONOMY

Male students appear to be singularly work-directed and able to function autonomously. They are in fact formally and, more importantly, informally very connected to each other, whether in the laboratory, in study groups, at conferences, on the basketball court, or in a bar. The male students receive informal 'mentoring' from male advisors who reflect themselves and see themselves reflected in these students. Even when men do not receive ideal support from their advisor interactions among peers and senior associates provide sufficient connection, feedback and information to shore up their self-confidence, thereby encouraging the capacity for assertiveness and risk-taking.

The existence of these mechanisms for support and connection belies the notion that somehow males are mysteriously constructed to be individualists, devoid of any relational needs. While acceptance of gender differences provides greater richness to complex questions, 'difference' does not indicate two rigidly distinct camps with no common ground between them. This notion gives license to a false perception of a fundamental disparity in personality structure, in which it is presumed that men are automatically programmed to function autonomously and women to be dependent. Such differential socialization is often falsely believed to decide, in advance, who can achieve in academic science and who cannot. But in fact, the way some male students function may not only reflect learned behaviors and coping mechanisms specific to gender, but also demonstrate the importance of identification with like others, based on subtle acts of inclusion and validation.

THE UNOFFICIAL PH.D. PROGRAM

An unofficial doctoral education process, based upon the establishment of informal ties, runs parallel to the official degree program of formal instruction, examinations and research production. Informal support structures and social gatherings provide information, encouragement and, most importantly, opportunities to learn from peers and role models in unpressured settings. Pick-up basketball games, pub visits with faculty, and study groups with fellow students to prepare for examinations are less open to women Ph.D. students than men in those disciplines where women have traditionally been scarce.

The induction of male graduate students into academic culture usually takes place with great ease. Knowledge is passed on about the informal rules of the game such as finding a compatible advisor and how to gain approval of a feasible thesis topic. This relatively invisible informal side of doctoral training in engineering and the sciences is more readily recognized in other disciplines. For example, it is well known that the motivation for attending a renowned business school to pursue the MBA degree stems not only from the cognitive content of the degree but also from the contacts to advance a future career that can be made during the course. Even when women are admitted to the official Ph.D. program, they are often still excluded from the unofficial, informal doctoral training process.

Women's precarious status has predictable social and psychological consequences that, if not countered, eventually affects scientific work. Particularly during the first two to three years of the Ph.D. program, women experience severe 'anomie' (loss of identity or meaning, a state of being without order), both psychological and social. In this context, a set of rigorous courses has the potential to challenge self-esteem built on earlier success. Feelings of 'anxiety–isolation–purposelessness' are the psychological counterpart to sociological anomie (Merton, 1938). In this instance, sociological anomie arises from the encounter with male-centered Ph.D. programs which disadvantage female students. A female graduate student described her predicament, 'I felt like I didn't have any back-up support. I didn't know how to pick a topic. The guys talk about that at the bars. I don't go there.'

The themes expressed by entering students revolved around the need to feel connected to others, to feel psychologically safe, to be given a professional identity, to be cared about, to be provided the strategies required to succeed and knowledge of the 'rules of the game'. However, the overwhelming experience of women is that of isolation and disconnection in their departments, and, in the most severely negative academic environments, among themselves. Thus, not only are they an alienated group within the department, they are isolated from each other as well.

Even when a woman was fairly well accepted, she was often excluded from crucial aspects of the graduate student experience. For example, a female doctoral student reported:

> We would all go to parties together and go and have beer on Friday, but if somebody came in to ask what drying agent to use to clean up THF, they would never ask me. It just wasn't something that would cross their minds. Nobody ever came in my office to ask what an answer was. People came in my office to ask the person who was in my room with me. I wouldn't have known if there were study groups . . .

The degree of invisibility of the informal education process is reinforced by a faculty informant who reported that through much of her graduate student career she was unaware that she was being left out of study groups; she simply didn't know that they existed.

EXCLUSION FROM STUDY GROUPS
The unoffical Ph.D. program begins with the formation of study groups

of peers, considered by students and teachers alike as the best way to prepare for doctoral qualifying examinations. This informal counterpart to the course structure consists of regular meetings of small groups of students in a department. Such groups provide a non-evaluative arena for thinking about complicated theories and articulating the jargon of the field. Technical knowledge is reinforced through discussion and informal presentations in these shadow structures to coursework and the qualifying examination. Departmental lore as well as other tacit knowledge is shared about faculty interests and idiosyncrasies that are likely to be transmuted either into exam questions or gaps in the examination regime. We identified some degree of exclusion from study groups in virtually all departments studied.

Participation in study groups and other social networks of peers in the department and the broader scientific community is an essential element of expected future success in science. An isolated individual has fewer intellectual possibilities. As one informant put it:

> If you're not in that scientific conversation then you're stifled. You can't get any help and you can't progress as far. Sitting and talking about scientific issues makes your brain work. Your creative juices flow and that didn't happen for me as a woman because discussions didn't occur. What was hard was that I was in class with all these people, and often getting better grades, and they knew I wasn't stupid, but it didn't matter. Oh, it was very isolating.

This exclusion from participation in study groups is not only personally painful but also removes access to a crucial component of graduate education.

QUALIFYING EXAMINATIONS

All students are concerned about qualifying examinations. However, women and men cope differently with this anxiety. Women tend to internalize difficulties and resort to self-blame, in contrast to men, who externalize and blame outside forces. Moreover, women are more likely to buy in to the likelihood that they will not pass. In the instance below, this student did not accept the self-fulfilling prophecy of a professor:

> [He] said, 'I think you should take them very soon so that you can fail them and then we can figure out what you need to do to pass.' I was struck that he expected me to fail, [that] someone can be that overt to me about their prejudices. I was able to go back to this person and say, 'You know, I was thinking about your strategy and what I prefer to do is figure out what I need to do to pass and then take them.' He actually became one of my biggest allies and was throwing questions at me once a week and I passed.

Another woman described her success in the candidacy examination as giving her a very strong push to complete the Ph.D. She felt that the experience raised her scientific maturity and provided ' . . . reassurance that I can complete a task.' Too often many women absorb the message that they cannot pass these examinations and elect to leave, particularly when they have failed once. A female graduate student said, 'I had very little expectation to pass and everyone had told me all along, you may get in, you can do the work, but you'll never get through those exams.' We suspect that the largest number of drop-outs may come either prior to the qualifying examinations, or even more likely, after one failure.

On the other hand, women who pass their qualifying examinations at the first attempt report a tremendous boost in self confidence. The successful experience with the examination is taken as proof that they will make it through the program. Often graded blindly, the qualifying examination comes closest to being a gender-neutral element in the Ph.D. program. Even when responsibility for its sections is handed over to a group of specialists in the field, the qualifying examination is a collective review. This, perhaps, explains why minor changes are always being made by professors in the department, 'especially in the qualifying exam'.

FINDING AN ADVISOR

Finding an advisor to work with is essential to attaining the Ph.D. degree. Students are expected to develop a close working relationship with their faculty advisor, a relationship that lasts several years and is crucial to the progress of the student through the program and out into the professional world. Yet entry into a relationship with an advisor is charged with ambivalence and ambiguity. Ph.D. students undergo a transition from a classroom to a research environment where they must learn to follow instructions and, virtually simultaneously, learn to make their own decisions. Thus, the content of the advisor–advisee relationship is likely to be even more significant, and more difficult, than the process of establishing the initial connection. Although it is presumed that the advisor has the most knowledge of the area of study, in reality the student soon accumulates a greater knowledge base in the particular area of their dissertation research. An imbalance between power and authority often emerges, in which near-absolute control rests with the advisor, even as the student's knowledge increases.

Despite the official existence of a committee for each student, most of the Ph.D. process is under the control of the individual advisor who has great leeway in defining the Ph.D. program for their students. The advisor decides what constitutes acceptable research for the dissertation and determines satisfactory progress.

A former student who had attained the Ph.D. discussed the necessity of developing strong ties with an advisor to reach that goal, especially given their discretionary authority. She said, 'One of the good and bad things about research universities is that the professors aren't really given guidelines . . . to turn students into scientists.' A female graduate student referred to the power of the advisor explaining, 'Most of the rewards come through the professor.' Despite the existence of a larger committee and even department-wide reviews of all students, there is great reliance on the opinion of the advisor. The advisor retains the authority to make the final judgment; the other professors on the committee are essentially there to support the advisor's decision.

The quality of the ongoing advisor–advisee relationship is crucial to the student's success. Difficulties in establishing a good relationship or deterioration of an existing one are signs of potential trouble in attaining the Ph.D. Without encouragement from a good advisor, a student can be lost and waste valuable time and effort. There is a great strain in having an advisor who is unable to provide guidance or shows a lack of concern with a student's progress.

For example, a physics student attributed her lack of direction to an inattentive advisor who was difficult to arrange to see, owing to a busy schedule. Their relationship deteriorated and the lack of contact contributed to the slow pace of her work. Even though she had found an advisor, the relationship did not provide the assistance needed.

Advisors have virtually complete freedom to make their own decisions, with the expectation that they will be supported by their colleagues. This situation can sustain both successful advisor–advisee collaborations, or contribute to a breakdown of relations without likelihood of repair. A student who developed a good relationship with her advisor used the advisor's help to plot a course which made the transition into research seem less abstract.

She explained ' . . . You have to know what you will be doing', and she described a 'settling in' process, a 'transition within the transition' in which, ' . . . as soon as I figured out what I wanted to do, I was happy with the work.' Having been appropriately guided, she did not 'float', the term some women applied to their state of lack of advisorial direction and support. A transition with a positive resolution left this student engrossed in research and finding satisfaction in her work.

Negative interactional patterns between male advisors and their female graduate students have been identified that, ' . . . [lessen] their opportunity for advancement' (Fox, 1988: 226). We also found a series of gender-related blockages to successful advising. Sometimes, there was an attempt at equal treatment based upon the faulty assumption that women had been socialized and educated the same as men. At worst, women graduate students were stereotyped as less capable and uncompetitive and were viewed as non-scientists. Such advisors simply could not take women seriously as graduate students. On the other hand, some male advisors were markedly more successful with their female advisees than some women faculty members.

We identified female experiences with male advisors ranging from the denigrating to the supportive. On the negative side are interactions that leave women with doubt about their self-worth. Even though this advisor probably thought that he was allaying concerns, the effect was the reverse. 'He said to me, "You don't have anything to worry about, they want women; so you'll pass [the qualifying exams]." You have the feeling, "Am I here because I'm a woman or because I am qualified?" It's like they take away all your achievements.' Women also discussed specific incidents in which their gender led to presumptions of lack of scientific ability.

For example, a female student was talking to a professor about her research problems and he said she was an 'emotional female'. She recalled, 'I couldn't believe he was thinking that. Maybe he was thinking I shouldn't be in physics. I always thought he was a nice guy. That's when I feel it: I'm out there on my own.' Male faculty members can exacerbate or mitigate the effects of traditional female socialization, depending upon their awareness, sensitivity and political stance on sex roles.

Most women are not socialized to understand the political strategies necessary to advance within the academic system. Without an advisor who is willing to encourage and direct, women are often unable to puzzle out the strategies necessary to get through graduate school. Women report that the best advisors are encouraging, give concrete directions and show them the ropes.

A women faculty member called attention to women's relative lack of knowledge of how to negotiate the academic system, explaining that many women lacked a strategy to deal with the admissions process:

> What you're supposed to do is get a hold of the brochure and if you want to get in at least say that's what you want. The women don't seem to have grasped that . . . the men go down the list and say, I want to work with this professor for this reason, that professor for

> that reason . . . the females give me no indication that they have
> even looked at the brochure.

This female faculty member suffered a particular sense of conflict since, in her own graduate career, she had taken a pragmatic approach, putting aside her own intellectual interests until later and pursuing her professor's research to get the degree in good time.

Attempts to find an analogy to the traditional female role for women in the laboratory are part of the notion of science as a 'male milieu' in which women's presence is viewed as disruptive and threatening. A chemistry professor used analogies from cooking in his discussion with a female student. (In a Japanese laboratory, the female graduate student took over tea and coffee duties in the secretary's absence.) These 'degradation ceremonies' may be followed up by subtle and not so subtle attempts to eliminate the unwanted presence.

For example, one woman commented: 'When I was trying to get something to work, [my advisor] would come up to me and say, "Did you see it yet?" Everyday he would say, "Did you see it?" I should have stopped it, but sometimes it takes a long time to see what's going on. It was very humiliating.' It is not only male advisors' treatment of female students that affects their situation but also how male advisors instruct their male students to act toward women. A female graduate student said, 'I hear rumors about myself . . . being involved with somebody. [I heard that] a faculty member was advising his students that it might be interesting to have an affair with me.'

Of course, men have also served as successful advisors to women. An offset to viewing women as 'sex objects' can be found in the following instance of advice about how to negotiate the shoals of negative behavior toward women. A sensitive male advisor helped this student make future decisions based on the reality of being a woman within the field:

> His attitude toward women is very understanding, very
> supportive, without being condescending. He doesn't say 'I
> understand what's going on,' which is offensive because it's hard

for a man to understand what's going on. He doesn't bring these issues up, I bring them up. He is very politically aware. He'll say, 'Don't talk to—.' Sometimes [his advice] was because of sexism and sometimes because this person was an arrogant son of a bitch and sometimes because this is a good person, but is just not comfortable with women.

Thus, women and men faculty do not, simply by virtue of their gender, automatically make good or poor mentors for female students.

Faculty who make the best mentors are aware of the different experience of men and women in the Ph.D. education process. They buttress their female students against the 'slings and arrows' of outrageous treatment. Sometimes they are willing to advocate change, going against prevailing conservative academic ethos with respect to academic practices. Traditional academic training programs are usually strongly believed to be meritocratic, even when and sometimes because they discriminate! In the next chapter we discuss the different experience of women and men in a 'male-centric' academic system.

6   Women's (and men's) graduate experience in science

The overall picture is of a prevailing academic culture that provides inadequate direction and mentoring for women, thereby eroding their self-confidence. In the first years of the program, women Ph.D. students experience the entire range of disorientation delineated in the Srole anomie scale: (1) the perception that community leaders are indifferent to one's needs; (2) the perception that little can be accomplished in the society which is seen as unpredictable; (3) the perception that life-goals are receding from reach rather than being realized; (4) a sense of futility; and (5) the conviction that one cannot count on personal associates for social and psychological support.

In addition, the individual is left with the feeling that it is she who is to blame, and this exacts a severe psychic toll including doubts about competency that prevent the successful working through of problems as they arise. It is not surprising that half of the informants revealed having sought personal psychological counseling during this period.

Isolation also creates powerlessness, loneliness, and confusion which, in many cases, leads to dropping out. Reports by informants describe how isolation reduces the opportunity (a) to compare experiences through communication with others in the same situation, (b) to test the reality of their experiences to ascertain that difficulties are not based on personal deficit, (c) to reduce feelings of alienation and rejection in hostile, male-dominated labs, (d) to work through strategies to deal with discrimination by male advisors, (e) to experience peer support when advisory support is non-existent, (f) to gain information and practical advice regarding strategies to succeed within the program, (g) to build a professional network among female peers for future professional advancement, (h) to feel safe to have questions answered without being judged as stupid or inadequate, (i) to practice the necessary skills for future advancement (presenting papers, discussing science).

NEGATIVE CONSEQUENCES OF ACADEMIC CULTURE FOR WOMEN

The academic structure, rather than aiding the passage of qualified and competent women, actively discourages them. The tiny cuts and stigmatizing reproaches experienced in graduate school range from assumptions of devalued admission to simply not having one's comments in a research group meeting taken seriously, only to hear them accepted when repeated a few minutes later, in a more glib and deeper voice, by a male counterpart.

Despite a formal and even at times a strongly stated commitment to non-discriminatory treatment of women, discrimination can be manifested informally. For example, a female graduate student reported different treatment of men's and women's contributions. She said, 'In group meetings I get the sense that if a woman says something, "okay fine" and that's the end of that.' In contrast, the response to males would be enthusiastic. Frequently compliments and praise would be given for the thought. This graduate student even mentioned that a woman might make the same observation and be met with a dismissal while a male student would receive accolades for the thought. The devaluation of women's scientific contributions is widespread (Benjamin, 1991) and takes many forms, including crediting the male partner in scientific collaborations and ignoring the work of women (Scott, 1990).

In some instances women are devalued by not being included in professional events. A female graduate student reported that invisibility was imposed when 'you have a visitor to the lab, the professor introduces the male students, but does not introduce you.' Another reported self-imposed invisibility in reaction to expectations that her contributions would not be valued:

> [In lab meetings] you feel very self conscious saying what you
> think and I think it's because you are a woman. They would just as
> soon you would sit back and be quiet and when they ask you if it
> turned red or green, [you say] 'it turned red,' rather than saying 'it
> turned red and this is what we're going to do next.'

Made to feel uncomfortable, women sometimes hold back from contributing their ideas to the scientific give and take of research group meetings.

The graduate school experience, as constituted at present, is often counterproductive. It results in the loss of many brilliant female minds to science and creates damaged identities instead. How is a secure scientific identity created? A sense of competence is related to the esteem of others for one's contribution and is further enhanced by a feeling of acceptance and inclusion by others. This amplifies a sense of self and ultimately frees us to take chances. The AIDS researcher who, against well-accepted methodological practice, mixed several samples together in order to have sufficient material to conduct an experiment, exemplifies the scientist as risk taker (Haritos and Glassman, 1990). In this instance the risk paid off; had it failed the individual would have been subject to ridicule, embarrassment and the censure of colleagues. To take such a chance, and to be prepared to accept its negative consequences, requires a secure sense of self. Without it, such scientific risks are not likely to be taken.

Women Ph.D. candidates are frequently mystified and sometimes struggle with guilt as to why they feel unable to enjoy the psychological freedom to assert themselves and take similar risks to their male counterparts. However, to enjoy such freedom requires connective tissue in which two powerful needs are met: 'the striving for autonomy in which self-organizing, self-enhancing and self-determining needs may be freely sought, and the striving for harmony which is the need to relate to and feel a part of a larger whole' (Ullman, 1992). These universal needs are inextricably interwoven and interdependent on one another.

The development of autonomous functioning, highly valued in the scientific work ethic, cannot be accomplished without full membership and inclusion within the social psychological milieu of the scientific workplace. Isolated and without interpersonal connection, a woman's ability to be playfully creative is impeded. Moreover, she is understandably reluctant to ask for the help she needs since it is likely that she will then be labeled as 'dependent'. A gendered 'apartheid system' exists in which many male advisors offer support to male students, but leave women to figure things out for themselves. With no support or connection with an advisor, taking risks in the lab becomes too threatening. People only take risks when they feel safe to do so. In contrast, there is sufficient support and acceptance, by way of informal interactions with male advisors and peers, for male students to enjoy the freedom to be innovative.

Women found it difficult to be taken seriously as professionals outside the department as well. One said: 'If I go to conferences, if I ask a question, the answer gets addressed to a man in the room. It's worse in physics than in other fields.' A female graduate student reported her response to being ignored, 'It's always a thing where being invisible, you don't exist . . . It was in a sense, I didn't exist.' Other times, women are made to feel different by being made too conspicuous. A female graduate student reported that a professor was ' . . . addressing the class, "Gentlemen" . . . and then made a big pause and looked at me and added, "and lady". I was different. Other people noticed it . . . '

Still other times women are patronized. A female graduate student told how 'I was sitting at this table and he kept referring to us as "my girls." In that context I didn't like it. He was thinking of us differently. He didn't say "my boys."' At one department, many graduate women felt that they were treated as 'one of the boys' but this too was an unsatisfactory resolution. Since the demands or possibility of child-bearing were not taken into account in structuring work schedules and evaluation, women were placed at a disadvantage, nevertheless.

LIFE-COURSE EVENTS

Academic transition points sometimes coincide with life-course events that affect how decisions are made. For example, as mentioned earlier, a pregnancy that coincides with such critical transitions as finding an advisor will set a female Ph.D. student at a disadvantage, if decision makers view child-rearing and research as inherently incompatible. A female graduate student said: 'There are no real good role models to follow. The women a generation ahead of us had it so difficult that they are by and large a very aggressive group. [They had to be so aggressive] and that's who got ahead. You have trouble looking at them and saying, "I want to be like that." You don't.' Even as taken-for-granted academic practices continue to work against them, most women in science do not want to be 'men'. Instead, many attempt to legitimize a female model of doing science (*Science*, special issue, 1993).

Male expectations about female commitment to family roles often lead to further discrimination against women in academic science. Many scientists believe it to be legitimate to take family responsibilities into account in evaluating a colleague, irrespective of demonstrated achievement; this is held to be the converse of a commitment to long hours spent at the laboratory site which is positively interpreted, irrespective of how they are spent. A female junior faculty member reported:

> I asked [her mentor and colleague] what his reaction would be if I had a child. He said, none. Then he said, 'I take that back. There are others in this department who will say, "Well, she won't be around now." A decision to have a child before tenure will have an impact on your tenure decision.' He was always extremely supportive. It was devastating [that he did not understand]. He's somebody who has good politics, who has been supportive of women. It was shocking to me. That did play a big part in my decision to stop working with him. I have felt completely isolated since then.

The barriers discussed in the previous chapter are exacerbated by the desire of most women, and an increasing number of men, for a personal life beyond the work site and the inability of academic science to accommodate their wishes.

MARRIAGE AND FAMILY

Marriage and children negatively affect women's careers in academic science at three key times: having a child during graduate school, marriage at the point of seeking a job, and pregnancy prior to tenure. In addition, we found some disparagement of marriage during the graduate student career. Women, but not men, are sometimes thought to be less than serious about their science if they do not stay single while in graduate school. As a female graduate student recalled:

> When I first interviewed to come here, I was single. On my first day of walking into this department I had an engagement ring on my finger. [My advisor's] attitude was 'families and graduate programs don't go together very well.' First he was worried I was going to blow my first year planning my wedding. I got a lot of flack about that and so did other women . . . teasing. 'So and so's not going to get much work done this semester because she'll be planning her wedding.' [sarcastically] The guys don't plan weddings.

Earlier in the century, marriage was grounds for a woman's expected retirement from a faculty position. The mutual exclusion of academic and family life has a long history. Until well into the nineteenth century Oxbridge male academics were also expected to choose between academic career and marriage. Neveretheless, there have been few if any residual carryovers from the academic celibate role for men, whereas for women, even when a choice between academic career and family is no longer an offical requirement, the presumption that each role requires a woman's total attention survives. It next surfaces when children are contemplated or arrive.

Women graduate students expect that they will be penalized for having children. One informant visualized her advisor's and the department's reaction: 'If I had walked into —'s office and said I was pregnant, they would have been happy for me as a woman, but in their list of priorities as to . . . who to support I would have plummeted to the bottom of the list.' These concerns arise because the existing academic structure is ill equipped to deal with pregnancy. Pregnancy is discouraged and graduate women who have children are encouraged to take leaves of absence that tend to become permanent withdrawals. In one department an informant reported that: 'The only one left is — [of the students who have children]. Two women Ph.D.s who got pregnant were strongly encouraged to take leaves of absence. One did and one did not come back.' In another department a female graduate student reported:

> One person took a leave of absence to get married and asked her advisor if she had a child would she be able to work part time and he told her, 'Absolutely not. No way.' What if I should want to do something like that? Is it the end of my career in —? Was it just the advisor? What am I going to do with my life? People say they're not going to have children until they're 40 and have tenure. I can't think like that. Thinking about [these] details is what scares me. That's when I think I should drop out.

The expectation that women students will succumb to the pressures of child-bearing and child-rearing makes some male and female faculty members wary of taking on women students in the first place especially since funding is tight and every place must be made to count. Another female faculty member stated:

> If a student had a baby with her, I wouldn't have her. Students who have babies here get no work done. It's not that I wouldn't take a woman with a child in the first place, but the first sign of trouble, I would just tell them to go away. If my students fail it looks bad for me.

Graduate student women were caught in a bind, wanting to have children and, while doing so, wanting to show that they could keep up with the pace of graduate work. A female faculty member reported:

> I had one student who was having her child in the middle of the semester and was to take and pass her qualifiers at the end of the semester. She wanted to do it. I said, 'Don't do it' . . . because of the emotional state you are in and the physical state after having a baby. We discussed this at length at one of our meetings . . . she ended up not doing it.

One department had taken child-bearing into account to a limited extent:

> During evaluations, if a Ph.D. [student] has a child she will be given some leeway for that semester . . . I think that's pretty funny . . . it's such a small amount of time. I think the women should get more leeway, you're physically out of it. It should be longer . . . at least a year. What's the big deal? [In one case, a student] had the baby in November and had until the end of the semester. It was partly her fault as well; she did not want to say she could do less. The faculty gave her a choice of doing a part-time thing or keeping up to pace. She chose to be put to the same standard as everyone else.

A peer had a somewhat different view of the faculty's action and described an unusual instance of solidarity among women graduate students:

> She decided not to take a leave [when she had the child] and made the decision at the end of the semester when we are all evaluated. She got a particularly harsh letter, [the faculty] essentially threatened to cut her support. They gave her requirements that would not be achievable for anybody . . . even without a baby. Two people had left the department earlier in the semester. One was a new mother, the other was a man who was very involved with his family. We got the feeling this was being done to

discourage her and tell her to go away. She was encouraged by her husband and a number of us to renegotiate this because it was clearly off base and came out of the blue.

There is a strong cultural bias in most of the academic science departments we studied against women combining parenthood with a graduate career; most advisors expect students to delay having children until after the degree, but then, when is the 'right time' if a woman stays on the academic track?

THE ACCUMULATION OF DISADVANTAGE

Barriers to women deriving from the structure of the academic system are reinforced by 'cumulative disadvantage' factors that excluded other women from science but also carry over and affect the academic careers of those who persisted. Beyond cumulative disadvantage carried over from previous negative experiences lies the realm of 'marginal disadvantage', irritations, the tiny cuts and stigmatizing reproaches experienced in graduate school. Disadvantage experienced at the margin of presumed success, after admission to a prestigious graduate program, is the unkindest cut of all. The fall to failure from such a lofty height is brought about in many ways.

Cumulative disadvantage extends back to the differential socialization of men and women. Girls are encouraged to be good students in so far as they expect to be given a task, complete it well, and then receive a reward from an authority figure. The roots of this problem lie in the different experiences of boys and girls. As young girls and women, females are socialized to seek help and be help-givers rather than to be self-reliant or to function autonomously or competitively, as are boys. In graduate school, despite the underground support structure provided for male students, behavior is expected to be independent, strategic, and void of interpersonal support. These expectations are antithetical to traditional female socialization. In addition, the needs of women, based on socialization which encourages supportive interaction with teachers, is frowned upon by many male and some female academic staff as indicative of inability. As a female graduate student put it: 'The men have the attitude of "Why should people need their hands held?"'

Many women come into graduate programs in science with low self-confidence. Women in physics, chemistry, and computer science reported that their graduate school experience further eroded their confidence. A female graduate student described the following symptoms: 'Women couch their words with all these qualifiers [because they are so insecure] . . . "I'm not sure, but maybe . . ."' One female graduate student said: 'I have the symptoms of the insecure woman. A comment from a professor can cripple me. I would be self-deprecating. My science is different because of my socialization, not my gender.' Another woman reported, 'Women tend to measure themselves: "Am I allowed to do this? This I know and this I don't know. This I should be ashamed I don't know."' Depletion of confidence is a signal of impending disaster.

An insecure person is like a weakened immune system, vulnerable to destruction from even a mild attack. If things are working out well, then initial lack of self-confidence is not too important: but if problems arise, then negative feelings come forth. For example, one woman had this to say: 'It is much worse if a woman fails an exam because her self-confidence is so low. I got an A- on an exam and was upset. The man sitting next to me got a C and he said, "So what?"' Another woman described the invidious comparisons that she began to make if things were not going well: 'If I'm not feeling good about myself, I start comparing myself to these brilliant people [highly qualified foreign students]. It doesn't affect American males as much.'

Finally, if the barriers remain high, low self-confidence translates into an increased rate of attrition. This loss can be viewed as a result of the cumulating thwarting of the development of a viable professional identity. Even those who do not give up, or are not pushed out, often reduce their professional aspirations.

Young women who remain in science and engineering Ph.D. programs, as well as those who leave, frequently describe expending a great deal of emotional energy in order to cope with a harsh social environment. A woman who left a Ph.D. program in chemistry after investing three years of effort said:

> There is no impetus that [my family] can give me that I would put myself back in that situation. There was no feedback on how I was doing, no pat on the back for what I had done. No feeling that I could knock on the door to initiate that kind of conversation. And constantly living with sexist joke telling. It was a complete blow to my self-esteem for the first time in my life . . . I was always successful in finding summer employment in chemistry, winning internships, getting science scholarships. Then I came here and I couldn't survive.

Similar feelings were expressed by women who persisted to the degree despite the alienation they experienced. For some women, experiences of denigration, rejection and dismissal are sometimes so elusive that they are not recognized until years later.

When rejection inexplicably follows great success a person is ' . . . left feeling inadequate and a failure, particularly when an individual has, up until this point, held a different view of herself' (White, 1974). Such a cumulatively deprecating experience erodes one's sense of personal worth. The female chemist drop-out further elucidated the effect on her of leaving the Ph.D program:

> It was really the first failure. The first major failure. I still view myself as intelligent enough, hard working enough to have earned that paper. I guess part of me views my graduate experience as a big black mark on an otherwise successful life. I very much wanted to earn the Ph.D. This continues to be an open wound because I didn't finish.

The psychological toll of such an insidious experience has consequences for how one adapts to the situation or if one even chooses to remain a scientist.

Ironically, most women Ph.D. candidates view graduate school as just as stressful for their male peers as for themselves. They are perplexed as to why they lack the apparent self-confidence and assertiveness of their colleagues. Anxiety often escalates into self-blame, exacerbated by feelings of inadequacy. Women report feeling increasingly anxious, and careful, desiring more direction in their research, and quick to blame themselves for perceived failure. In contrast, they observe their male peers as more assertive, action-oriented and risk-taking. These behaviors are cited as evidence of 'independence' and 'autonomy', and that lack of these characteristics is frequently mentioned by an older generation of male scientists as the rationale for women's 'inherent' difficulties in academic science.

The findings discussed above have been corroborated elsewhere, for example in site visits to assess the climate for women in physics departments (Dresselhaus *et al.*, 1997) and in a study of three other science disciplines carried out in 1994 by the Association of Women In Science. There have also been a few attempts to supplement qualitative evidence by querying and comparing broad representative cross-sections of students of both genders. One such recent survey (Curtin *et al.*, 1997) which aimed at all female graduate students and a comparable number of randomly selected male students in physics in a given year, provided confirmation, albeit modest, of the picture that emerged from qualitative studies. Among students who were U.S. citizens, women students were somewhat less likely to describe the faculty as easy to discuss ideas with (38% as against 52% for men), or fellow students as respectful of the respondent's opinions (72% compared to 87% for men). U.S. women were also slightly more likely than men (15% to 8%) to indicate a currently unfulfilled wish to belong to a study group.

However, most other aspects of departmental life evoked only muted differences between male and female students. These included respondents' relationships with other students in their research group, their sense about whether other students in general treated them as colleagues, and the degree to which they are encouraged by faculty members. Indeed, most graduate students gave positive evaluations of their department environment, with a few notable exceptions such as whether the department encouraged student self-confidence, or whether department faculty as a whole treated students as colleagues, and such reservations issued more or less equally from male and female students alike.

One clue to understanding these apparently equivalent findings for each gender is buried in the comments provided by the respondents at the end of the questionnaire. Women who had given negative evaluations of their graduate experiences quite often elaborated specifically on a departmental climate that they felt was hostile to women, whereas male 'complainants' discussed issues such as the poor job market they faced once they graduated or the overall quality of their coursework and the coverage of their program's curriculum. Thus, although men and women seemingly evaluate the overall graduate environment similarly, women note distinctly different bad experiences. The human price for the Ph.D. is higher for women than for men, and the rewards are often lower.

A good graduate school experience can allow the effects of previous disadvantages to be left behind. Too often, old bad experiences interact with a new set, further lowering self-confidence. This concatenation of disadvantage, as it is disentangled, explains the cumulative thwarting of female scientific talent. In conjunction with lack of a viable professional identity that should have been nurtured in graduate school, it produces reduced aspirations. A male faculty member said of his female students, 'Their job aspirations are so low, their self-confidence is so low, they tend not to apply for what they see as a very tough place.' The effects of traditional female socialization are exacerbated by the assumption that women should fit in to a 'male' academic culture, instead of that culture being reformulated to accommodate both sexes.

INFORMAL TRANSITIONS: THE ROLE OF CONFERENCES
A key hopeful finding is the identification of participation in conferences as a significant informal transition point. In addition to

'vertical' transitions through the stages of a Ph.D. program, there are also 'lateral' transitions in which the student moves out of the research group and department and into the broader scientific community. If Ph.D. students participate in conferences, it widens their social circles and allows them to envision their future in the scientific community. Several respondents brought up the topic of conferences without being asked and discussed how their participation had enhanced their graduate career. One advisor's suggestion to a student that she take part in a conference was taken by her as a signal of his high regard. Here we see exemplified the role of the advisor in assisting their graduate students and moving them forward. A female graduate student pointed out that 'not just everyone can attend . . .' and that the invitation gave her a feeling that she was 'doing the right thing'; that she was ' . . . on the right track.'

Conferences thus play an unexpected role in the socialization of female scientists, providing information and social support that might not otherwise have been available. Since women experience problems at various points in the Ph.D. career, a transition point that provides a positive experience takes on a greater import for women than men. Participation in conferences builds confidence and gives women a chance to network on a new level. A female graduate student explained that ' . . . being sent to conferences happens in accordance to your relationship with your advisor, specifically it depends on how you please your advisor.' Conferences give additional support to those students who have proven themselves capable of doing excellent work; those who are invited or permitted to attend are pushed into an environment that allows them to make connections they would otherwise not find.

Perhaps the most important event, at sub-specialty workshops and conferences, is being introduced to key senior scientists and fellow graduate students from other departments. While larger meetings are widely publicized, smaller meetings and workshops are often by invitation only. In any event, it is typically the student's advisor who can insure that the student gains the maximum benefit from participation. Legitimation from the advisor, through a few words added to a personal introduction about the quality and potential of a student's research, means that they will be taken seriously from the outset by peers. It is through these introductions that the advisor's social capital is placed like a mantle around the student, guaranteeing that whatever she does or says will be taken seriously.

Invitations to speak at or simply attend conferences are especially important to women in furthering their graduate careers. At least three positive effects can be identified: (1) an increase in the female student's confidence from a favourable reaction to a research presentation; (2) introduction into scientific networks, paving the way for future conference invitations, job possibilities and research collaborations; and (3) reinforcement of the advisor–advisee relationship, as both parties recognize its place in a broader social network. The more recognition they received in the scientific world beyond the department, the greater the acceptance female Ph.D. students felt in their home department.

FINISHING THE PROGRAM

Issues of isolation, lack of direction and contacts, and conflict around one's life chances continue to dominate toward the end of the program. A sixth-year student admitted that even though she had only six months left before finishing, she frequently considered seeking counseling. She reported feeling overwhelmed with anxiety about the future and obtaining a job even though she had spent five years in industry before entering the program: 'I was feeling left out. I didn't know where I belonged. The longer I have continued in this work [the more I have felt], "Where am I?" If you're not feeling good, your self-confidence is going down . . . and on top of that you have no money and going in debt, I think that's another consideration [to make you feel like quitting.]'

The only reports of women who elected to drop out toward the end of their graduate school career concerned those who apparently had earlier despaired of remaining in their science owing to difficulties within the department. After having negotiated continuous conflicts with either advisor or committee, the candidates finally decided to remove themselves completely from a rejecting and distasteful situation by withdrawing regardless of the consequences for their degree.

Not feeling 'cared about' is thematic throughout these interviews. There is frustration that there is no group or individual geared to meet the needs of upper classwomen. For the most part, these women have banded together but find themselves alienated from the mainstream, with little access to learning 'the rules' and gaining access to 'the club.' Advanced female students found that male peers belittled their accomplishments. Male student's attitudes typically reflect what filters down from the male faculty, a complacent, dismissive denial of women's scientific ability.

Many advanced female students were not struggling with issues regarding their dissertations or finishing their degrees. The women who had reached this point had been able to locate an advisor-advocate; those who dropped out had not. Their paramount concerns were for the future, after graduation, 'how their lives were going to go.' They wished to find someone to ask about negotiating a balance between employment and family. Lastly, it was at this juncture, when they were close to the completion of the Ph.D., that many realized that they were devoid of professional contacts and networks as they sought post-doctoral fellowships and employment, and this struck home. Those with children were now concerned about career choices and finding jobs that would allow for time with their children.

POST-DOCTORAL FELLOWS

A recent study of female post-doctoral fellows concluded that, with the possible exception of biology, men in positions of power (doctoral and post-doctoral advisors, tenured professors) often harmed women's scientific careers, intentionally and unintentionally (Sonnert and Holton, 1996). Invidious differences arose from male professors not taking women seriously as scientists; although present in their labs they might not do something as simple as passing on information about a relevant conference or as crucial as using their connections to further a promising woman's career as they would for a man. One woman even reported not being introduced by her sponsor to a laboratory visitor, a seemingly surprising but all too typical instance of discrimination also reported in other studies of research groups.

Gender constraints from the larger society further reinforced the male culture of science which tended to make women 'invisible.' A significant number of women reported that they selected a fellowship site not on professional grounds, but in order to follow a husband whose career had priority. Each of these disadvantages might individually appear to be a small matter, an oversight or a matter of personal choice. Yet, over time, advantages and disadvantages accumulate; more often for men into a 'Matthew effect', the halo of success that attracts additional rewards and renown (Merton, 1968), but for women into a 'Cinderella effect' where the reverse conditions hold.

Several Ph.D. students in our sample had graduated and moved on to their post-doctoral placements. Some who had good relations with their advisor at the graduate level reported poor experiences at the post-doctoral level. A post-doc at an Ivy League university said that she received no supervision at the start of the fellowship. Although her situation has since improved, her transition was an isolating experience. She was left alone in her work; nobody noticed her presence. Writing her first paper by herself was a difficult task in these circumstances. She believes that supervision would have helped. In retrospect, she decided that she had received considerable moral support during graduate school, especially in comparison to her post-doctoral experience. She hadn't realized at the time how good it was.

Another student working on her post-doc in New Mexico reported that the work was similar to thesis work she had completed, the only difference being that she is now more highly paid. A student pursuing her post-doc at a cancer research facility in Philadelphia described the environment as being a good hybrid, in between academia and industry, and a good transition, especially if one was interested in a career in industry. Both of these fellows had taken the initiative in obtaining their positions, either through networking at conferences or simply writing a letter to the chair of a department. But these are not typical cases. Instead of a rich experience of ever-increasing integration into a scientific community, as women move to higher levels, many report isolating experiences.

OVERCOMING THE EFFECTS OF ISOLATION

Few women who attain advanced degrees acquire the density of connections that typically accrue to men as they move into the academic system. Increasing the flow of women through the pipeline by removing blockages to entry and exit from Ph.D courses of study is a necessary but insufficient policy. To attain the maximum value from investment in human capital, it is necessary to recognize that the quality of women's Ph.D. experience is as important as the numbers of degrees granted to women.

The crucial relationship for Ph.D. students is with their advisor; the second most important is with fellow students. Female graduate students report problems with both male and female advisors. Feelings of incompetence, self-blame, isolation and confusion arise from poor relationships with advisors. Without the support of an authority figure, women consistently reported feeling lost and incompetent. Early in their graduate school experience, they were often unable to gain their advisor's attention and support. Later in the degree program some reported compensating support from peers that helped them persist to the degree. Under conditions of relative isolation, attainment of a Ph.D. degree could be merely a formal achievement, lacking the penumbra of informal connections that arise from being introduced into a scientific community by a mentor.

Student participation in conferences was identified as a critical informal transition point. Conferences, in addition to providing a forum for the dissemination of scientific knowledge, are also a venue for distribution of 'social capital', the connections and access to information and resources that help build a research career. However, full participation rests on much more than simply receiving funds to attend. Although access to resources is important, it is the advisor's introduction of the student to colleagues that is crucial in forming relationships that will be important to future scientific success. Women report mixed experiences in being introduced into the broader scientific community in their field by their advisor.

Conducting regular 'body counts' of Ph.D. production, by sub-discipline and department, is an important first step toward evaluating graduate education in the sciences. 'Quality of academic life' indicators should also be constructed. A female graduate student referred to transition points as 'threats', suggesting that intimidation is still the norm. The continuing perception of transitions as dangerous appeared to contradict indications that her Ph.D. program was moving away from a 'weeding-out' approach.

Even though changes have been made, the previous system, or at least its image, is still intact. Critical transitions for women in science are not yet 'rites of passage' into a welcoming community; instead, they are often fraught with peril for female scientific careers. As women ascend the educational ladder, they increasingly find support at the early stages, only to later encounter the exercise of arbitrary authority or simple inattention to women's needs.

WOMEN'S (AND MEN'S) GRADUATE EXPERIENCE: A SUMMARY

Getting a Ph.D. involves far more than passing qualifying examinations and producing high quality research for a dissertation. Success in graduate school is highly dependent upon being included in the informal social relations of academic departments. Even though men and women are in the same graduate programs their experience can be strikingly different. Most men quickly become included in the informal aspects of departmental life while women are often left out. No matter how brilliant and academically successful an individual has been in the past, isolation can take a toll.

The academically superior women in our study, who had typically been at the top of their school and undergraduate classes, were shocked upon entering graduate school to find themselves marginalized and isolated. They were often excluded from study groups and left to grapple with course work and examinations on their own. Many either found themselves deterred from attaining the Ph.D. or received the formal diploma without becoming part of the social networks that are an important prerequisite for future scientific accomplishment.

All of the female students interviewed for this study had had highly successful undergraduate careers. Most reported strong mentoring relationships with a special advisor, professor, or lab director (usually male) who recognized their scientific potential and encouraged them to apply to graduate school. As one respondent put it, 'I think all of us were very successful [before coming], otherwise we would not be here today. I was very successful, cruising through my undergraduate classes.' She continued, 'The one thing that really made me decide to go to grad school was the experience of doing one-on-one research with a professor at my undergraduate institution. He gave me a lot of encouragement which gave me a lot of confidence, all of which has been drained since the first month here.' Although a disproportionate number of women are deterred from graduate training by discouraging experiences in college, the smaller number that do go on have typically had a superior experience that all too often is not repeated at the next level.

Ideally, an educational institution should provide for ongoing development, with each succeeding stage providing new opportunities to further consolidate and advance past achievements. U.S. elementary and secondary education has a mixed reputation. Universities, however, are viewed as the crowning glory and saving grace of an otherwise flawed educational system. Because of this aura of exceptionalism, graduate school is usually examined as a unique closed system: socially, legally, and dynamically different even from other elite institutions. Nevertheless, graduate school is a social-psychological milieu like any other place of work. It can therefore become either a source for self-realization or a place where interpersonal interactions are blocked (Berger, 1967). In this chapter our focus has been on the experiences of women in graduate programs in the sciences and engineering and the effects of both formal and informal structures on their training. The different graduate educational experiences of females and males make it almost seem as if there are two Ph.D. programs in the same department: one track for men, the other for women.

The female 'track' has long been less well populated. In recent years an increasing number of women have pursued higher degrees in the sciences, bringing the tension between women's lives and the taken for granted 'male' structure of the Ph.D. program to light. In an earlier era when women in science were very few in number, unconscious negative effects of a training system that did not take women's interests into account were all but invisible.

The fact of a growing population of women at the higher levels of scientific education does not tell the whole story of what is within those advancing numbers. In the next chapter we address the anomaly of dispersion and isolation within this numerical increase: the paradox of 'critical mass' for women in science.

7 The paradox of critical mass for women in science

At each transitional point the number of women decreases at a significantly higher rate than men. Thus, while women made up 37% of the students taking physics in U.S. high schools in 1988, only 22% of those taking the calculus-based introductory physics course in college were women (AIP, 1988, 1991). Women's presence is reduced to 15% of those receiving the bachelor's degree in physics and then to 10% of the share of Ph.D.s. The decline continues in the shift from education to academic employment, with women constituting 7% of assistant professors of physics and only about 3% of full professors.

What are the effects of such small numbers on the women who persist in scientific careers? A key finding in our interviews was that as the number of women faculty members in a department increased, they divided into distinct subgroups that could be at odds with each other. Senior female scientists typically shared the values and workstyles of older men; their narrow focus failed to meet the needs of most younger women. In contrast, some younger women (and a few men) struggled to create an alternative scientific role, balancing work and non-work issues. The scientific role thus divides along generational and gender fault lines. These developments have significant unintended consequences for the socialization of female scientists – for example, the availability of relevant role models. As long as the relatively few women in academic science were willing to accept the strictures of a workplace organized on the assumption of a social and emotional support structure provided to the male scientist by an unpaid full-time housewife or done without, issues of women in science were not attended to. A modest increase in the numbers of women in science, without a change in the structure of the scientific workplace, creates a paradox of critical mass.

Affirmative action is expected to clear up blockages in the pipeline but many of these barriers persist. Affirmative action rests, in part, on the premise that a sufficient number of persons from a previously excluded social category is required – a critical mass – in order to foster the inclusion of others from that background. From the 1970s, efforts to increase the number of women in academic science departments have largely resided in affirmative action programs, requiring full consideration of female and minority candidates. However, in the 1980s lack of vigorous enforcement reduced the spirit of the law into a bureaucratic requirement that became a routine part of the paperwork of the academic hiring process, often with little or no effect on recruitment (Nuevo Kerr, 1993.)

A minority group (especially one that has traditionally been discriminated against) is easily marginalized when only a small presence in a larger population; its continued presence and survival is in constant jeopardy, requiring outside intervention and assistance to prevent extinction. As the group's presence and level of participation grows, at a particular point the perspective of members of the minority group and the character of relations between minority and majority changes qualitatively. In theory, the minority is increasingly able to organize itself and insure its survival from within and effects a transition to an accepted presence, without external assistance, in a self-sustaining process. The discrete point at which the presence of a sufficient number brings about qualitative improvement in conditions and accelerates the dynamics of change is known as 'critical mass'.

The magical statistic for a critical mass has sometimes been defined as a 'strong minority' of at least 15%. This implies that there are a sufficient number to have an impact on the majority. But as we shall see in this chapter, it matters what the 15% represents: if it represents women of several nationalities who self-isolate within their own communities, leaving two American women in two unrelated laboratories unknown to each other, it is a false number. It is also a false number if the departmental culture is so toxic that the freedom to associate with other women is subtly restricted. It is also a false number if the 15% is so dispersed within multiple laboratories that it is not known that there are other women with whom to interact. The outcome in all these scenarios is the perpetuation of isolation. So the precise number is less important than the nature of the response that the new minority receives from the majority – in this case, female scientists from their male counterparts.

Critical mass, the presence of a significant minority whose precise number varies by context, has contradictory effects. Indeed, as underlying conditions improve, the situation of the minority group may appear worse as formerly repressed grievances come out into the open. For example, sociologist Paula Rayman and co-workers (1996) have reported that as a 'critical mass' of women appears in medical school, the rate of sexual harassment cases increases. One explanation is that the rise of empowerment is at work, with women feeling safe enough to file complaints, as well as there being more women available to be subject to harassment!

When women are a token minority they may well fear the adverse consequences of raising controversial issues and complaints. As their numbers increase and they become better organized, women are more likely to avail themselves of redress procedures. At one university that we studied, the affirmative action officer reported an active caseload in the humanities and social sciences but virtually none in the sciences and engineering. She attributed this difference to the 'universalistic' character of science.

However, in our interviews we found women in science and engineering departments with similar complaints. With token numbers and an unorganized presence, these women expected highly undesirable consequences, placing their degrees at risk, if they brought an action. In contrast to the humanities and social sciences, science and engineering departments lacked an organized political support structure at the time of the study, leading women to repress their grievances and giving their departments a false appearance of gender equity.

In one instance, a woman graduate student at this university contemplated making a complaint against a male faculty member who was discussing pornographic images on a computer screen with his male graduate student. The incident took place in her presence in an office that she shared with the graduate student. She drew back from making an official complaint, fearful of endangering her degree. However, the matter became so widely known within the department that the chair sent out a strongly worded message condemning the practice as unacceptable and warning against its repetition. The chair's response to this incident, which received some publicity on the departmental and nationwide e-mail networks for women in computer science, was an isolated event; the department remained basically unchanged in its treatment of female graduate students.

Change without struggle is less likely than conflict with determined resistance. Under certain conditions, an organizational transformation culminates in minority group members achieving and retaining positions of real power and authority that were previously beyond their grasp. Paradoxically, ostracism of women often accompanies the breaching of gender uniformity, the first stage in the breakdown of resistance to women's participation in science. The initial reaction of men to the appearance of women in a scientific field, research unit or academic department has typically been to ignore them.

When they must be taken seriously because of demonstrated accomplishment, there is often a negative reaction couched as a criticism of a woman's personality or appearance. The fear of ostracism often leads some women in science who have 'made it' to deny the existence of the gender-related obstacles in their path. Calling attention to difficulties that they have managed to overcome could lead to counter-charges that they received special privileges and thereby devalue their achievements. Younger women are often concerned that participating in activities for women in their department will set them apart; they perceive that men will look askance at this affiliation and are sometimes unwilling to participate as a result.

This finding implies that the minority must attain power to overcome resistance, as opposed to findings that a modest increase in numbers, by itself, results in improved conditions. The support of persons in structural positions of power, or attainment of such positions by members of the minority group, is the key to change. In this view, despite accretion of numbers, it is strategic power that really counts. Despite their apparent success within the existing system, women faculty members struggle with feelings of inadequacy regardless of their status.

As one female faculty member summed up the condition of women in science, 'I guess it's our socialization. I have a lack of self-confidence myself. I guess I've gotten more confident as I get older and take on more jobs like editor in chief of a journal and so on. I still notice feelings of lack of confidence and maybe I'm not good enough to do this. I see it in lots and lots of my colleagues. I see it at the faculty level and especially in young graduate students.' This phenomenon explains why many women, especially junior faculty members, do not feel that they can afford, either socially or professionally, to be activists or advocates for young women students within their departments.

ALTERNATIVE THEORETICAL PERSPECTIVES

Two theoretical frameworks have been offered by sociologists to explain the effects of tokenism: the 'group interaction' perspective and the 'demographic group power' perspective. The group interaction literature indicates that women suffer in work groups where they are present in small numbers. Kanter (1977) argued that minority group members or 'tokens' are less likely to be accepted by members of the majority, a process that she called 'boundary heightening'. In addition, tokens face increased visibility and pressure to perform that negatively affects working conditions and reduces performance. Social contact with majority members is lessened and when such contact occurs it is often based on stereotypes. Significant differences appear as the ratio of minority to majority members improves.

Spangler, Gordon and Popkin (1978) found that women students in a law program composed of 20% women earned lower grades, tended to select 'ghettos' law specialties and participated less in class than women in a law school with 30% women students. Alexander and Thoits (1985), Gutek (1985) and Konrad (1986) found similar effects on grade point averages, relations between male and female co-workers, and self-evaluation of work. The group interaction perspective suggests that as the proportion of minority members in a group increases, achievement levels improve but not necessarily in a continuous fashion.

Alternatively, the demographic group power thesis argues that a subgroup's ability to gain organizational resources is proportional to its size. Thus, some group power theorists argue that the higher the proportion of women, the greater their ability to improve their share in the distribution of resources. However, while the increase in the size of a discriminated-against sub-group may improve members' ability to influence the distribution of rewards in their favor, it also engenders increased resistance from majority members who expect to suffer corresponding losses.

Indeed, South *et al.* (1982) found that among office workers, as the proportion of women increased, their interaction with male co-workers and the support they received for promotion from male co-workers decreased. Thus, the larger the minority the greater the discrimination against it, causing the culture and experience in different departments to seem impervious to incremental change.

IS CRITICAL MASS SUFFICIENT?

Attainment of critical mass only partly resolves the dilemma of women in academic departments and research units. The fallacy of critical mass as a unilateral change strategy is shown by the fact that female faculty and senior research scientists pursue strikingly different strategies. Despite some progress, organizational structures within units, and the divisions they engender, continue to isolate women.

Furthermore, the dispersal of women students into male-dominated research groups sustains isolation even when there is a potential critical mass in academic departments. Nor does an improvement in the total number of women in a unit necessarily overcome an underlying situation of subfield fragmentation that further increased the isolation of women.

Quite often a department is divided into subdisciplinary groupings that have little to do with each other. For example, one chemistry department in our sample was so fragmented by area that neither a female faculty member respondent nor the students interviewed had an accurate picture of how many women graduate students were in the department. This fragmentation seems to be especially pervasive in chemistry, more so than in other disciplines studied. When the numbers are low to begin with, once in research groups, women students often do not see one another, and occasionally barely know of each other's existence.

In this environment, small numbers have two implications. First, members of any group characterized by small numbers will have a statistically lower chance of being central figures in different networks. Second, although critical mass is a prerequisite for access into powerful social networks it is not enough on its own, because social networks criss-cross sub-specialties in the field, research topics and geography. Nevertheless, we also found that a modest increase in the number of women did bring about some change in departments.

In this respect, critical mass does work smoothly: there is more support and safety in numbers. A female student observed, 'One good thing is that there were female faculty members. It definitely changes the attitude of how male students react to women. They must take them seriously and this is positive.' When senior females were present, overt male behavior toward women improved (for example, public sexual joking and stereotyping declined) in a threshold effect of critical mass.

THE EMERGENCE OF FEMALE-GENDERED SUBFIELDS

Women who are in positions of power can act as mentors to help bring other women into central networks, rather than leaving them in minor subfields where women traditionally congregate. Of course, what is central and what is peripheral changes over time. We are currently in the midst of a transformation in which the male-dominated physical sciences are being displaced from their scientific and economic central position by biological sciences, both molecular and evolutionary. Women have traditionally made up a larger proportion of biologically related fields.

The concentration of women in subfields is not always voluntary; women are also subtly or not so subtly directed to these fields. It is significant that these areas are usually biologically focused. In chemistry at one university, biochemistry had a significant number of women and the area was connected to the bio-science division. In computer science, many women gravitate to artificial intelligence (AI) where cognitive processes and psychological links are prevalent. In electrical engineering, the bio-electrical subfield attracted the most women.

Biology has evolved sufficiently in its gender composition and character into a field where attitudes toward women have changed. Many informants expressed this as a given: when they were told that we were looking at biology, their assumption was that this field was less problem-ridden owing to the numbers of women present. There is a snowball effect: as the numbers increase in the biologically related areas in electrical engineering, chemistry and computer science, they then attract still more women.

The difficulty in relying on creation of a critical mass of women faculty members in a department as a change strategy is that the faculty members may be unable or unwilling to play the designated role. Nevertheless, even though it is not the whole solution, a significant presence of women on the faculty allows students to become acquainted with female role models, even if some are persons they do not wish to emulate. As one faculty woman put it, 'Women are attracting women. The broadcasting of women-friendly environments through the grapevine [is a step forward]. Women are being empowered as women are increasingly a scarce commodity of departments seeking female faculty.' Beyond the creation of women's networks is the issue of opening up all networks to female participation.

8   The 'kula ring' of scientific success

Elite male participants in a cultural complex extending from the Kwakiutl Native American communities in the Pacific Northwest to the Trobriand Islands in the South Pacific regularly meet to give away their most prized possessions to each other. The more material goods an individual gives away, the higher their social status and the more secure their standing within the group. This pattern of social relations, called the 'kula ring' in Melanesia and the 'potlatch' in North America, provides an informal means of organizing and redistributing resources and power in the community (Drucker and Heizer, 1967). 'And it is just through this exchange, through their being constantly within reach and the object of competitive desire, through being the means of arousing envy and conferring social distinction and renown, that these objects attain their high value' (Malinowski, 1922: 511).

Conducted at regular intervals, the kula ring has its analog in the way scientists exchange ideas, resources, and information. Like gatherings of Melanesian clan leaders, elite scientists who are linked by ongoing networks of relations and governed by norms of trust and reciprocity ritualistically meet to discuss collaborations, discover complementary areas of research, and introduce their graduate students and post-doctoral fellows to each other for future correspondence and employment.

Even members of the same department may experience the scientific world quite differently, depending on the configuration of their social networks. The differential effect of networks on women and men has also been noted, particularly in regard to problems of mentorship and networking across sub-areas that differ in status (Ibarra and Smith-Lovin, 1997). Suzanne Brainard, director of the Women in Engineering Initiative at the University of Washington (an organization that promotes the inclusion of women in the engineering professions) concluded that 'The major issue facing women at the academic level is isolation' (Emmett, 1992).

Epstein (1970) found that in medicine, law, and science, men get more opportunities for professional advancement through informal sponsors who provide advice and share tacit knowledge on how to get ahead. Reskin (1978) documents this exclusionary process for women in scientific careers. She found that when there are few women in an academic department or in an industrial or government research unit, and women are not well accepted by men in those settings, they experience the effects of isolation.

Post-graduate women in academic research groups also experience professional isolation more acutely than men as they become aware that they are not being invited, to the same degree as their male peers, to be part of the professional network that leads to contacts and potential job openings. A female graduate student commented on the frustration that this kind of perception creates about career prospects and job aspirations and said, 'If you had a job, who would you hire first? Someone you're buddies with, right?'

The benefits of being in a strong network of contacts are the mirror image of the problems of isolation. Early inclusion in a strong network provides a 'jump start' to a scientific career. For example, a professor's invitation to a graduate student to deliver a paper at an elite conference allows network building among fellow graduate students and their senior sponsors. Such connections create a stable and supportive reference group as well as providing channels by which to disseminate work and share ideas.

In this chapter, we argue that differences in scientists' social networks influence their career success by shaping their level of social capital. Like the more familiar concepts of human capital (a person's talents and know-how) and financial capital, social capital has exchange value and can be accumulated. It is different in that it depends on relationships to create and sustain it.

SOCIAL AND OTHER CAPITALS: DEFINING OUR TERMS

Social capital is one of several forms of 'capital' that have recently been recognized by analogy to monetary capital. Thus, human capital or 'what you know' is the intellectual reservoir of ideas, methods, and factual knowledge that one accumulates, whereas social capital or 'who you know' is the web of contacts and relationships that provide information, validation, and encouragement. Social capital refers to the relational aspects or informal dimensions (e.g., commitment and intimacy between persons) while 'cognitive capital' is the knowledge base that an individual has acquired in a particular field. Both kinds of capital, human and social, in optimum combinations are the key to achievement, reward and recognition in science.[1]

Thus, social capital refers to the productive resources a person gains access to through contacts that control critical resources, or creates with another person they have a relationship with, but which decrease in value if the relationship ends or the resources are transferred to another relationship. Social capital 'accrue[s] to an individual or a group by virtue of possessing a durable network of more or less institutionalized relationships [that are] embedded in a stable system of contacts possessed by an individual' (Bourdieu and Wacquant, 1992: 119). These resources include interpersonal trust and norms of

---

[1] The 'social capital' thesis had its origin in a debate over the relative importance of family ties or formal schooling in occupational success. The social capital literature focuses on the role of the home and the school in reducing or mediating the negative effects of social origins through provision of a network of trust or what Granovetter (1985) calls 'embeddedness'. 'Social capital' also refers to the nature of relations between individuals based on their mutual expectations and obligations, channels of communications and social norms (Coleman, 1988). Families at all economic levels are becoming increasingly ill equipped to provide the setting that schools are designed to complement and augment in preparing the next generation. Within the family, the growth in 'human capital' is extensive, as reflected by the increased levels of educational attainment. But social capital (relationships and social ties), as reflected by the presence of adults in the home, and the range of exchange between parents and children about academic, social, economic, and personal matters, has declined, at the same time that the parents' human capital (e.g. level of education) has grown.

reciprocity as well as knowledge of new scientific ideas and strategies for developing a line of research.

Social capital provides an approach for analyzing differences in the success of men and women in a social context in which productivity is based on managing interdependence with others. Our hypothesis is that the role of social capital increases in a non-linear fashion. As individuals attain initial increments of social capital, their likelihood of obtaining future infusions grows ever greater. The presence or lack of connections to a mentor or role model of scientific success gives some individuals a head start and places others at a disadvantage. Importantly, social capital accumulation is all too often gender linked and makes a difference even among the successful. A recent Danish study of access to post-doctoral fellowships found that women had to have 2.6 times the publication productivity of males to achieve equal career success. Our hypothesis is that a greater measure of human capital was required to make up for a deficit of social capital.

We begin by describing how the production of scientific knowledge has changed from solitary work to production lab work that places new emphasis on social networks as the mechanism for linking interdependent scientists across departments and universities. Next, we describe the creation of social capital in scientific careers and its effect on both careers and knowledge production. This material provides the background for a later chapter which empirically examines the differences in the social capital of men and women scientists and the ways in which it influences their aspirations and career chances. Then we describe the role that social capital played in the race to discover the structure of DNA. Finally, we draw a series of inferences from the data analyzed earlier in the chapter about the relationship among women, social capital and science.

THE NEW ORDER OF SCIENTIFIC PRODUCTION

The model of scientific practice in the natural sciences that was once taken for granted, and was expounded by Vannevar Bush in his landmark vision *Science: The Endless Frontier* (1945), has materially changed. Bush argued that science follows three simple linear steps – basic research, applied research and development – and that a rich human capital base and heavy investments by the government drove the success of scientific achievement. The important factors are two: a strong, government-led financial infrastructure and a strong physical infrastructure composed of state-of-the-art technology, laboratories, instruments, and facilities. In the fifty years since *The Endless Frontier* the myth of the individual scientist has been supplanted by the reality that the development of scientific innovations requires not only financial and human capital, but also a store of social capital.

Nevertheless, the myth of the individual scientist persists as an instance of what anthropologists call 'cultural lag', where a belief about a practice may continue to be strongly held long after reality has irrevocably changed. Thus, funding agencies and scientists themselves refer to the 'individual investigator', who is actually a research manager, whether in academia or industry, who provides intellectual inspiration for the work of several graduate students, post-doctoral fellows, undergraduates and technicians, and who also raises the funds from granting agencies to support their work.

The managerial responsibilities assumed by the contemporary academic scientist who conducts funded research lead many of them to say that they feel like they are running a small business . Indeed, these academic quasi-firms have many of the characteristics of a business firm, save for the profit motive (Etzkowitz and Peters, 1991). Certainly academic science has come a long way organizationally from the individual faculty member before the Second World War, who might work with a few graduate students individually. The contemporary academic scene requires a continuing replenishment of resources, of varying kinds, to maintain the viability of a research group.

The pivotal role of social capital over and above the effects of financial and human capital is illustrated by recent research on successful university/industry partnerships in Great Britain, Germany, and the U.S. from 1850 to 1914 (Murmann, 1998; Murmann and Laudau, 1998). Murmann carefully measured and compared the government expenditures on R&D and education in these nations, as well as the quality of research faculty and the number of Ph.D.s awarded. He demonstrated that the decisive factor for the advancement of science was not the quality of the financial or human capital (even though some base level was obviously needed), because these were at equal levels in all the countries studied.

Rather, the nature of social ties among scientists in the academic community and industry was the key factor that stimulated scientific innovation and the likelihood that laboratories, led by university professors, would develop world class research and products. It was the higher level of social capital among German scientists – that is, their ties within the university, and between university and industry – that enabled them to pool resources and ideas more effectively than those in other nations.

We believe that social capital is even more important in scientific production today than it was nearly 100 years ago. First, beyond the formal structure of courses and examinations, contemporary advanced scientific training is increasingly based on transfer of tacit knowledge in group settings among peers and with mentors and through learning-by-doing.

Second, most problems are too large for a single individual to tackle alone, given the increased specialization of subfields and the high costs of analysis, testing, and product development. The success of research projects increasingly depends upon the coordination of several scientists and non-scientific groups that bring diverse scientific as well as managerial and financial competencies to bear on problems that are increasingly at the interface between science and commerce.

The complex reality of rapidly developing fields, in which knowledge is both sophisticated and widely dispersed, demand a range of intellectual and scientific skills that far exceed the capabilities of any single organization, as illustrated by two notable recent discoveries in biotechnology. The development of an animal model for Alzheimer's disease appeared in a report (*Nature*, Feb. 9, 1995) co-authored by 34 scientists affiliated with two new biotechnology companies, one established pharmaceutical firm, a leading research university, a federal research laboratory, and a non-profit research institute.

Similarly, a publication identifying a strong candidate for the gene determining susceptibility to breast and ovarian cancer (*Science*, Oct. 7, 1994) featured 45 co-authors drawn from a biotechnology firm, a U.S. medical school, a Canadian medical school, an established pharmaceutical company, and a government research laboratory. More important than the number of authors is the diversity of sources of innovation and the wide range of different organizations involved in these breakthrough publications (Powell *et al.*, 1996:118–119). All of this work suggests that as knowledge becomes more fragmented and dispersed among individuals and organizations that are geographically separated and institutionally independent, the role of social networks in forming information bridges and providing social support increases.

THE MOBILIZATION OF SOCIAL CAPITAL IN SCIENTIFIC CAREERS

Social networks emerge as a matter of necessity around scientists who must pool resources and talent, providing members of these networks with special access to information that is not available to those outside. An important aspect of social networks is that they influence how individuals exchange ideas, information, and resources in ways that other forms of information transfer, such as journals or gossip, do not. In particular, a quid pro quo logic of exchange is the norm in networks as new members (e.g., incoming graduate students) feel indebted to others in the network for bringing them in and sharing with them their knowledge and resources. As this indebtedness builds up among different members in the network, the social network becomes a repository for opportunities and investments that can be drawn on when needed and are reciprocated at some time in the future back to the group.

In this sense, networks share similarities with revolving credit associations – associations in which members donate resources to the group and draw on them when in need. The result is that social capital builds at different rates depending on the structure and membership of a person's network, and that social capital is created as a by-product of many different activities which share a common thread of generating, consuming, and investing in productive resources.

As this process extends over time and to new contacts, an individual's social network is enriched with opportunities (social capital) that can facilitate success. Contacts can be called to access information about jobs before they are nationally posted, to discover what areas might receive special funding consideration, to locate unpublished material, or to garner the validity of new research findings. Being invited to participate in research grant proposals by senior faculty members increases chances of success; exclusion sends a strong message to women to seek another position before tenure review.

In one department that we studied, junior female faculty members in an electrical engineering department reported that they were left out of invitations given to young males to participate in large-scale research proposals. The negative message this gave to one female faculty member was so strong that she resigned her research position in favor of a job at a teaching college. This capital also plays a critical role when evaluation is equivocal (e.g., the evaluation of work), when favorable interpretations are important, or when referrals are made to individuals in other networks with other opportunities for jobs, research projects, and funding. To begin a career without these connections isolates an individual at the very stage of a scientific career when visible achievement is crucial to long-term success (Merton, 1968; Lin, Ensel and Vaughn, 1981; Fernandez and Weinberg, 1997).

Senior scientists who have developed a high degree of social capital can be thought of as 'social capital bankers' and typically have the greatest access to information; their networks are widely cast, cross-cutting research teams, departments, and universities (Zuckerman and Merton, 1972; Seashore et al., 1989). These individuals are in the best position to spot opportunities and to connect less embedded individuals to opportunities. Persons new to departments, such as graduate students or assistant professors, are in the opposite position. They lack a large and well-developed network of opportunities and therefore are highly dependent, for information and contacts, upon individuals in central positions.

Elite scientists deposit social capital with their protégés and fellows in the form of invitations to high-level conferences and access to privileged information. Participation in a circle of investigators of related topics gives a scientist confidence to move swiftly in interpreting his or her findings, increasing the likelihood of being credited for a new discovery. Information gained through telephoning friends can be a crucial part of preparing an experiment and assessing results. Such ties provide informal access to knowledge about emerging topics for investigation along fast-moving research fronts, a hint of related results, or warnings about unfruitful approaches already attempted elsewhere, so-called 'telephone science.'

Offering access to protégés, who are viewed as rising stars enhances the prestige of a senior scientist and further expands their power just as the ceremonious circle of gift giving, among the Kwakiutl tribe, produces the 'social glue' that keeps the dispersed clans together as a collective entity. As with the Kwakiutl, access to participation in informal ties and networks is the surest path to high achievement and enhanced social standing in the group (Burt, 1992).

Exclusion from a network of social ties and critical mass contributes to the well-documented decline in the proportion of women who make it to each succeeding rung on the professional career ladders (Burt, 1997). But the exclusion also affects those seemingly 'successful' women who manage to persist all the way to the Ph.D. Here the issue is not one of survival, but of the quality of the experience in graduate school, and the efficacy of the tools that are available to make the transition to a productive and satisfying career.

Attaining formal status such as an advanced degree or a research position is a necessary but not sufficient condition for a highly successful career in science. Formal positions are only a rough indicator of success, since individuals of the same rank differ widely in the strength of their networks and their access to scientists with relevant knowledge for possible collaboration.

Thus, social capital plays an important role in enabling scientists to manage the interdependencies inherent in scientific labor and practice. It does not replace human capital or financial capital as the primary force but reveals the importance of resources that are lodged in relationships between people rather than in individuals, technology, or institutional systems in scientific careers and work. Social capital increases an individual's timing, access, and referral benefits, which span the contexts in which other 'capitals' operate, and also links them together in new combinations and innovations that result in the creations of privileged resources (Burt, 1992). While we postpone to a later chapter our exposition of how social capital is measured in the context of the hard sciences, we note that it depends on the structure of a person's network and the quality of the relationships they maintain with people who occupy that structure.

Large expansive networks of weak ties offer access to new and novel information while strong ties provide social support and a basis for gaining social and political support (Granovetter, 1973; Coleman, 1990; Burt, 1992; Uzzi, 1997). Our hypothesis is that men's and women's different career trajectories can in part be explained by looking at their differences in social capital.

SOCIAL CAPITAL, DNA, AND GENDER

In his memoir on the discovery of the structure of DNA, James Watson described the epiphany that he experienced at an international poliomyelitis congress in Copenhagen. After a week of ' . . . receptions, dinners and trips to waterfront bars . . . an *important truth* was slowly entering my head: a scientist's life might be interesting socially as well as intellectually' (Watson, 1968: 40; italics added). Sharon Traweek, an anthropologist who conducted a participant observation study as a public information officer at the Stanford Linear Accelerator Center, observed the same kind of social interaction noted by Watson but among the average scientist and away from the spotlight of fame. At this leading high-energy physics research site, she found that groups of men from each detector congregated at lunch in the cafeteria and after hours at local pubs to build friendships and exchange ideas. The same was true of theoretical physicists, who, unlike experimental physicists, were expected to be 'solo artists' because they experienced few technological or financial incentives that might divide the labor of a project among a workgroup of fellow scientists.

Watson apart, the informal social relations of science usually go unreported in the history of scientific achievement. The news sections of *Science* and *Nature* do not yet have a gossip columnist to chronicle who was seen with whom at a meeting or to note new collaborative partnerships. Ironically, a solitary existence has been assumed to be the mode of scientific life. Consequently, an examination of the effects of isolation on scientific careers goes against the grain of popular expectations that 'scientists work alone, in silence' (Traweek, 1988: 57).

Nevertheless, news is routinely transmitted among colleagues, locally and at a distance through informal channels, third-person gossip, phone, and e-mail. There are times, such as when two groups make simultaneous discoveries, when gossip among scientists is intensely conspicuous. But ties among scientists not only permit information flow, they increase the probability of finding scientific partners with complementary skills, a shared commitment to a particular line of work, and the conviction to stand by their recommendations and convictions.

Consider the case of Rosalind Franklin, Watson and Crick, and the discovery of DNA's helical structure. Why wasn't Rosalind Franklin included among the recipients of the Nobel prize for this achievement? Owing to her untimely death, the inevitability of her exclusion can never be proven. Rosalind Franklin produced the first photographs containing the crucial evidence of DNA's helical structure. But she worked in an isolated research environment apart from the assistance of a single junior colleague.

Franklin lacked membership in a group of colleagues who might have supported riskier inferences and encouraged her to publish her findings more fully and quickly. Instead, in the prototypical fashion of an isolated female scientist, she built up her battery of evidence slowly and precisely before she was willing to draw strong conclusions. She lacked the social capital needed to locate persons with particular scientific information that was outside her expertise that could have allayed her hesitation about the speculative nature of her findings. In addition, she lacked contacts to researchers who could have mobilized support for her insights.

In contrast, her competitors Watson and Crick set forth ill-supported hypotheses on the chance of hitting the mark, despite the risk of being shot down, because of the support furnished by their large and well-connected network. After announcing a theory that was quickly found to be empirically wrong, they were ordered to cease and desist from this line of work by the Institute's head. Nevertheless, Watson and Crick obtained informal support and expressions of interest from colleagues and soon geared up for another assault on their goal.

Watson and Crick also used informal social interaction with scientists outside their new laboratory to complement the knowledge that was shared among scientists and technicians in their own laboratory. For example, the chance presence of a structural chemist at the Cavendish laboratory helped confirm the validity of Watson and Crick's helical model by revealing an error in another published work that was the foundation of Watson and Crick's insight. This was the same information that had eluded Franklin.

Watson reported that the discovery was due in part to the 'unforeseen dividend of having Jerry share an office with Francis, Peter and me.' He also noted the opposite effect on one of his competitors, a scientist working on the same problem: ' . . . in a lab devoid of structural chemists, [he] did not have anyone to tell him that all the textbook pictures were wrong. But for Jerry, only Pauling would have been likely to make the right choice and stick by its consequences.' Had it not been for the network ties with chemists at the Cambridge pub, the misunderstandings of the chemical structure induced by the inaccurate pictures in chemistry textbooks might have never been uncovered – or at least not in so timely a fashion. Similarly, the large and diverse network of personal contacts possessed by Watson and Crick enabled them to mobilize resources and to effect a quick recovery after an initial failure.

A related example of the effects the lack of social capital is seen through the eyes of Evelyn Fox-Keller and in her analysis of Barbara McClintock, a future Nobel Laureate who, like Watson, was associated with the Cold Spring Harbor Laboratory. Evelyn Fox-Keller's graduate school experience in physics in the late 1950s and early 1960s shows the dark side of networks – isolation. She described how a lack of a supportive network negates motivation and lowers levels of aspiration in ways that undermine an individual's ability to realize their potential at a crucial point in the maturation of a scientist (Shapiro and Henry).

Fox-Keller almost dropped out of her Ph.D. program after being systematically excluded from participation in the informal social relations of her department. 'She was making a name for herself indeed, but the process was a nightmare. She was completely isolated . . . She was going to get out of physics' (Horning, 1993). Fox-Keller told her interviewer, 'My real world began to resemble a paranoid delusion. Many people in Cambridge knew who I was and speculated about me. None of them offered friendship.' After two years of such treatment, she 'was a wreck – defensive, weepy and unapproachable. She passed her orals but decided not to do a thesis.' Fox-Keller eventually completed her doctorate after a revivifying visit with biologists at Cold Spring Harbor Laboratory whose 'attitude to her was a lot more accepting than that of most members of the Harvard physics department' (Horning, 1993).

Her own experience of isolation turned Fox-Keller's scientific career from physics to molecular biology and eventually to the history of science. Years later, preparing for a colloquium, Fox-Keller experienced the converse of Watson's epiphany, recalling to consciousness the effects of her isolation. She had repressed her painful experiences, including the haunting image of a distinguished female scientist seen on solitary walks at Cold Spring Harbor. Fox-Keller wrote, 'Barbara McClintock represented everything that I was most afraid of – that becoming a scientist would mean I'd be alone . . .' (Horning, 1993). This realization led Fox-Keller to refocus her career on the analysis of gender issues in science, starting with a biography of Barbara McClintock, a Nobel prizewinner who shared with Watson an interest in genetics and for a time a research home – the Cold Spring Harbor Laboratory.

And what she showed was that despite being Nobelists and having exceptional scientific minds, the career trajectories and networks of McClintock and Watson were vastly dissimilar. McClintock was an outsider who, on the fringe of key networks of communication and support, operated at a competitive disadvantage relative to Watson. While a few persons like Barbara McClintock turn social capital disadvantages to advantage (Fox-Keller, 1980), most fall behind in professional attainment or decide to change careers.

RAISING SOCIAL CAPITAL
In this chapter, we have described the nature of social capital, how it forms, and the ways in which it can influence scientific achievement. We drew an analogy between social capital and the kula ring. Both are defined by membership, the special logic of exchange that governs resource trades within the group, and the revolving 'credit' and investment nature of the exchanges that tends to enrich the network over time.

We placed social capital in context by reviewing literature that showed how the change in the way scientific work and practice is conducted – from graduate school recruitment to the conceptualization and marketing of scientific goods – revolves around interdependencies among scientists that are best managed through social capital. Our point was not to discount the importance of human, cognitive, or financial capital in career success. Rather it is to show the effects of social capital, particularly in how it enables scientists to find collaborators with complementary skills and resources and to invest in public goods that are available to members of their network.

Social capital develops out of interaction with prior contacts and network structure in several ways.

First, reciprocity governs exchanges between contacts. New faculty members who are invited into a network feel indebted to the others in the network for bringing them into a supportive and status-enhancing social group.

Second, the new members' initial indebtedness and reciprocity expand as members exclusively share tacit knowledge on how to get ahead or allocate discretionary resources to each other. For example, contacts can help one to find out about jobs, pass on information about research areas that might receive special funding consideration, locate unpublished material, or make introductions/referrals between formerly unacquainted persons who share similar interests (Kenney, 1986; Seashore *et al.*, 1989; Powell *et al.*, 1998).

Third, the right social network configurations not only get access to but also increase the speed and veracity of the information transferred. These outcomes are also important for the university. In decentralized systems, non-bureaucratic systems, informal connections provide a by-way by which information is accumulated and imported from other institutions (Powell *et al.*, 1998).

Fourth, networks provide social support to their members. Close working relationships with experienced network members help new members interpret critical feedback and motivate commitment to a long-term program of scientific study, which is often punctuated by few immediate rewards. Emotional support and group affiliation create an identity that enhances feelings of self-worth ('I know others feel the same thing'), generate commitment to goals that have delayed pay-offs (Ibarra, 1992), and provide a group mechanism for legitimating claims or counteracting against discrimination (White, 1992; Podolny and Baron, 1997). As an individual's portfolio of contacts grows over time, resources and opportunities also accumulate, facilitating success (Coleman, 1988).

Scientists at central positions in scientific networks function as repositories for social capital because they have a large number of connections to diverse persons in and outside the department. Those at central positions in a network can be thought of as 'social capital banks' which accrue gifts or loans from other senior scientists with similar reservoirs of relationships. The members of the network gain special access to information that is not available to non-network members, and to information before it becomes available to the 'market', namely scientists outside the network.

Moreover, these connections facilitate the match between a person's human capital (i.e., a person's specific skills, talents, or motivation) and various situations in which opportunities appear to use and expand one's capital. Thus, the network is the organizing structure of interpersonal ties in which social capital accrues and, depending on their network access, individuals with equal human capital experience different rates of success based on their social capital (Coleman, 1988; Burt, 1992; Ibarra and Smith-Lovin, 1997).

In the next few chapters, we turn to an in-depth analysis of the experiences of women faculty and the ways in which different departmental structures affect their ability to succeed in science and for science.

9 Women's faculty experience

Despite the continuing existence of barriers to women, a generational change in the traditional 'male model' of full-time devotion to science and neglect of personal life is under way. A senior female scientist in an academic department has often been an individual, successful by conventional measures, who chose to adopt the strategy of emulating the 'male model' as the only way to survive. Treated as 'one of the boys,' she often later has second thoughts about the sacrifices that had to be made to be accepted.

A decade ago, we identified a small number of women faculty members who were limiting their time in the laboratory and attempting to integrate a private sphere with their professional life. Recently, more women as well as an increasing number of younger male faculty members have expressed interest in a less-driven work life but stringency in research funding has intensified the pressure to work more. Even though some report that their satisfaction has decreased under these conditions, the most driven scientists submit an increased number of grant proposals and become even more successful. The conflict between their behavior and the wish to change suggests that transition to a more equal balance between professional and personal life is still a long way from being realized, especially at the higher levels of academic science.

Until quite recently relatively few women were willing to openly articulate the vicissitudes of their professional and personal experience in science. Aware of the responsibility their status carries as role model and trail blazer for younger generations, they have not wanted to inadvertently discourage the aspirations of their potential successors in a scientific career. Moreover, needing to safeguard the validity of their own personal achievement, based on impersonal criteria of merit, they have not wanted to 'rock the boat' by suggesting that scientific achievement is affected, either positively or negatively, by personal or social factors. The growing willingness, during the past decade, to discuss less than optimal experiences and concomitant ambivalence is an important indicator of change among female scientists who face a series of ambivalent situations in most science and engineering departments.

Wanting to encourage younger women to pursue scientific careers, successful women scientists are in a bind. If they focus upon gender barriers and funding difficulties, young women might become discouraged and respond very rationally that it is not worth the effort to pursue such a difficult career. To examine these issues, we conducted more than 400 in-depth, in-person, interviews (followed in some cases by re-interviews over the Internet), in several waves, with female faculty in twenty-one departments in five disciplines (chemistry, biology, computer science, physics, and electrical engineering) in both public and private research universities located in all major regions of the U.S.

On the surface, women who have attained faculty appointments at prestigious research universities appear to have crossed a significant threshold of status and achievement to a place where gender 'no longer matters' (Sonnert and Holton, 1996). These women scientists have become icons to a scientific community that proclaims the ideal of universalism (Merton, 1942), and to a society in flux around issues of gender. For most women scientists who achieve faculty status, full professional self-definition in the face of subtle and overt exclusion remains problematic.

In many departments it is dangerous to identify oneself with women's issues or other women. Even when there were two women faculty members in a department, they would typically stay apart from each other and not form a friendship until after tenure. Often lacking anyone to talk to, younger women discussed the conflicts between work and personal life in confidential interviews for this study that often took on the emotional intensity of a therapy session. In this

WOMEN'S FACULTY EXPERIENCE 133

chapter we discuss the experience of women faculty, in both negative and positive settings, to explain the troubled, and troubling, experience of women in academic science.

THE CASCADE EFFECT

Impediments to women in science appear at all stages and phases of the scientific career line and can be viewed as a 'cascade effect.' Like a series of interconnected circuits, the first member of the chain supplies power to the second, the third and so on. A cascade of affirming experiences serves to amplify a string of positive effects, until there is a short-circuit and the process is reversed. Women who have avoided negative experiences at an earlier stage often encounter them later. The majority of girls and women do not experience such uninterrupted multiple positive experiences as they ascend the educational ladder. Instead, at some point in time what had the potential for a cumulative positive cascade of experience becomes short-circuited by negative experiences.

Successful women in science view themselves as having had prolonged relatively positive experiences and attribute their status and achievement to supportive mentors along the way. However, the value of these early experiences becomes at risk when negative experiences begin to accumulate on a faculty level. Many are understandably unsure, especially before the tenure decision, how much risk they can afford to take in acting as mentors and advocates for their women students. Even after a permanent position has been achieved, the complex and, at times, contradictory experience of the successful woman scientist carries with it an aura of taboo. The path to faculty status begins with the job search, but social capital and gender strongly affect the outcome.

THE JOB SEARCH

We have identified several patterns of impediment at the point of the job search: a dominant one of deferring to a male partner, a less usual one of ignoring personal considerations. Departments also typically receive job applicants differentially by gender, taking into account women's personal obligations in deciding to hire while ignoring them for men. The assumption, of course, is that women will be strongly affected by their ties; men less so. A certain persona, strict adherence to a rigid academic career path and total time commitment are among the unstated requirements for many jobs.

Women who survive the strain of lack of support for child-bearing and child-rearing in academia and complete their degrees at the highest levels of achievement may nevertheless find that their career will not survive the next hurdle of the academic career path. When a married woman is about to attain the Ph.D., the 'two body' problem comes into play, typically deflecting women's careers from their highest potential. Two shifts in work site are typically needed: from Ph.D. program to post-doctoral position in a different university and from post-doc to yet another work site. The highest climbers on the academic ladder of success are able to accept the most promising and prestigious post-doctoral and faculty positions without regard to any other consideration. The rule of intellectual exogamy has disastrous career consequences for women who are unable or unwilling to make individualistic decisions on where to work. As one observer put it: 'The academic market is a national one. Those who do not accommodate their choice of geographical location and willingness to move to their careers may lose out' (Rosenfeld, 1984: 99).

Marriage and children are generally viewed by male faculty members as impediments to a scientific career for women. Even those most supportive of women take this view to some extent, as the following quote shows: 'I've had some disappointments with very good women who settled for jobs that are less than an equivalent man would do. You have some extremely good people you think are going to go out and make a mark and then somehow or other they marry somebody and spend their time in a bad career. For a man to decide not to take his career seriously is like admitting he takes drugs. For a woman to say she puts her family ahead of her career is considred a virtue; the pressures are all in that direction. The women are told, "Isn't this

wonderful. You are giving up your career to sacrifice for your husband."
The pressures come from society, relatives, to some extent the men
involved, the parents of the husband.'

On the other hand, a few women take a different tack. They are
willing to break off personal relationships that interfere with accepting
the best possible job. A male professor portrayed the situation of a
woman, involved with a man, who, he said, ' . . . could have gone either
way. I asked her, "To what extent is his career going to interact with
what you do?" She said, "Not at all. I want to find the best job I can and
if it works out for him O.K. and if it doesn't well then that's the end of
the relationship." So she had decided that career is what really
mattered. She's at [prestigious Eastern university] and he's still out in
California so that's the end of him. She took what I would say is a
typically man's approach to things, that the career is the primary
decision but they don't all do that.'

Women who enter the academic job search often find that they have
made a career detour that is held against them despite evidence of solid
achievement. A male faculty member expressed amazement that a
respected research institute had hired a female Ph.D. who had
temporarily left the academic track to become an astronaut. An
academic career gap for more prosaic reasons of child-birth and child-
rearing is officially expected to be ignored but is inevitably taken into
account to a woman's detriment. Affirmative action procedures have
too often been turned into an elaborate ritual of seeking out female
candidates for interviews and then determining that they cannot be
hired when compared with men who have followed the straight and
narrow path to academic success. If 'best' is defined in terms of an
aggressive persona, with numbers of publications the primary criteria
of achievement, then women will typically be defined as inferior.

GEOGRAPHICAL MOBILITY BARRIERS
The limited geographical mobility of many women restricts their
choice of both graduate school and job. A highly successful female
scientist interviewed in another study explained the impact of location

on her career, given existing norms of hiring. A research associate, her advance in rank was limited, as was her exposure to students and the experience of raising her own funds. She felt that these consequences of having to accept a position of lesser status had delayed her professional maturation. 'I was married – I'm still married – and I didn't have the flexibility of moving around. That's one of the best ways to achieve a permanent position and to increase one's standing; to have the lever or the threat of saying, well, I'm going to leave. And to mean it. You can't do it as an empty threat. You have to be ready to leave, and people are. I was never in that position, so I could never use that threat' (Dupree, 1991: 117).

A typical scenario that has been identified is marrying a man in the same field who completes his graduate work before his wife. He finds the best job he can without geographical constraints. When the woman finishes, she finds what job she can in a circumscribed region (Max, 1982). Women who are already married often select their graduate school based on what is available in a region and choose a job with similar considerations in mind. Second-rank research universities attract many higher quality candidates than they might otherwise, because of women's geographical restrictions. Of those of our graduate student informants who aspired to academia, most were interested in jobs in small teaching colleges rather than research universities because, as one woman summed it up, 'Science isn't everything.'

10 Dual male and female worlds of science

Two scientific worlds, one male, the other female, emerge from the faculty interviews. These dualities were expressed in the key moments of an academic science career such as setting up a lab and preparing for the tenure decision. The question of how to act as an advisor to female graduate students was also fraught with tension. With respect to these various issues, the female experience in academic science was typically far more difficult in contrast to the well-connected male. Lacking a satisfactory conduit for information, a junior faculty woman reported that she and her peers would 'sit and discuss for hours and hours what to do, then we walk away not knowing if we should do it because we are too young. We are brand new in this department and we don't know if that is the way to do it.' Even having peers did not necessarily help, for they were equally clueless.

For others, the separate and unequal experience was one of invisibility and denied professional identity. A new woman faculty member may be mistaken for a secretary, student, or technician by hostile older men, or considered inauthentic: 'I really don't know what they think [about me] because I interact with them so rarely. I mean I'm the only woman among 42 faculty members, so they don't know what to make of me, period. Most of the faculty here are used to treating women as wives and secretaries, or both.' Bereft of connection, there is an inability to check on day to day tasks as basic as 'people to talk to so that I could compare labs', and no mechanism for reassurance which would occur 'if everyone had a chance to talk to somebody. Then it would be okay. They could see things can work out.' Without such colleagueship, one's science is depleted.

The impact of a first experience of professional and personal ostracism is heightened because it is unanticipated at this level of achievement. Echoing the astonishment of many women graduate students who suddenly find themselves socially excluded upon entry into a Ph.D. program, this successful woman scientist describes her awareness of her own social emptiness when she reflects that 'I'm not on bad terms with these people. I'm on no terms. On every superficial level I have something in common with them, but I have no relationships. Everywhere I've been in my life, I've made friends. Here it is a black hole.' Another faculty woman reflects that 'I was never aware of any gender-based discrimination when I was a graduate student or post-doc. It's something I'm only aware of now that I am interacting with an older generation . . . I'm just coming to grips with the fact that there really is a problem. I've now seen enough people to know there is a problem.'

Neglect by colleagues can also hurt the development of a sense of professional identity. For the first time, reflects this junior woman, 'you begin to notice things you probably never noticed . . . I started the same time as my husband and his office partner. We're all about the same age. Everybody [on faculty] kept coming up to me when I first got here asking "oh, so who are you working for?" That never happened to those guys. These colleagues probably feel more comfortable, so they ask the guys to do things. So it gets lonely. A lot lonelier as you move up because you have no peers to talk to.' After being introduced at a party to welcome new faculty, a woman faculty member recounted that 'one organic chemist said to me, "Oh, I really respect women in science. You know women just aren't mechanically inclined, so I'm really impressed when I see one in science." At first you think it's an isolated incident and then it happens time after time after time, and you realize these people have a problem.' Women often face the constriction of supportive professional networks and collaborations, as helpful to achievement in science as in other endeavors.

Exclusion is also experienced as devaluing and undermining female faculty members' opportunities for scientific success. After a brilliant graduate school experience, one young electrical engineering professor recalled the feelings of abandonment and anxiety during her first year in the department while she attempted to relocate, set up her lab and teach: 'The department treated me like shit. The one guy who brought me up here, he didn't do a damn thing. It took eight months before my name showed up in the front. No one ever told me I could get a phone charge card, so for two years I've been paying for the calls myself. Before I got here the former chair said, "We'll put you up for a PYI" [Presidential Young Investigator]. I had been here three days and he said, "Maybe we won't put you up for it this year. Go out and get some teaching experience, go to some conferences and get to know some people in your field." Nobody told me about something I was responsible for until a week before it was due. It was my first proposal. They never appointed a mentor . . . a man or woman, it wouldn't have mattered. At the end of my first year I got extremely depressed. It was very bad. It took a hell of a lot out of me. I was very anxious.' Not only was the scientific potential of this young faculty member harmed, the department hurt itself by not encouraging all its members to achieve and thus advance the collective academic reputation.

Without full membership in the scientific community, a deprecating sense of self-consciousness appears to permeate the female scientific experience. Not only are the professional linkages missing for basic information and career building, but feelings of alienation give rise to vigilance, guardedness against rebuff and the need to 'prove oneself.' The outcome is a reluctance to 'climb the ladder to get something I want. I try to do it on my own. On my own capabilities,' as this woman biologist described. Ultimately, lack of full membership in scientific activity creates uncertainty and self-protectiveness. A protective response sometimes takes the form of niche work or perfectionism, finding an out-of-the-way research field to cultivate on one's own with such a degree of certainty that it could not be subject to attack. This is the obverse strategy of identifying a 'hot topic' that leading figures in a field cluster around simultaneously in a race to be first with a brilliant hypothesis or a definitive finding.

Since women faculty constitute only a tiny minority in many departments, generational discrepancies in values and scientific style among women themselves can further isolation. A young female faculty member expressed the feelings of many of her contemporaries whom we interviewed, 'One of my biggest problems here is gender bias from the older faculty. I never have worries like that from people of my own age. I really think it is a generational problem.' A female chemist who had a close collegial relationship with her advisor in graduate school, a contemporary, was unexpectedly stunned when she spoke in a faculty meeting: 'I got this guy so riled up that he sent a memo around saying it was totally inappropriate for junior faculty to speak at faculty meetings.' Experiences such as these help explain why the minuscule number of an earlier generation of women may have been left with no alternative but to adopt a style that would make them more acceptable to their male colleagues.

The effects of negative treatment of female scientists carry over from within the university to the external professional world. Women, for example, become reluctant to introduce themselves into informal groups at scientific conferences and meetings. A tenured biologist still feels ' . . . very isolated. It starts to cycle in: I start to withdraw, the more withdrawn, the more isolated. It gets more and more difficult to be there. My perception has been that it's a boy's club. It's hard to break in on them, especially if you are a young woman. My male friend says, "It's hard for me too," but then I point out to him that you may feel awkward, but you always walk over and when you do they separate, and there you are. I can't even walk over and I don't know what would happen if I did.' Isolation begets expectations of isolation in a spiraling fashion.

Even as they are discriminated against, female scientists are expected to assume the official responsibilities of minority status in academia. Women will be asked to take on more tasks within and outside the department than their male equals, because of their status as 'the token woman'. Paradoxically, this role enjoins that an individual who is invisible because she is different, become visible because she is different. It is, of course, an additional stressor in what is already an inordinately stressful new situation: 'When I first got here I was asked to be on a lot of university committees and [male] colleagues at the same level weren't asked. So I realized pretty quickly that it was because I was the only woman in the chemistry department and a lot of these committees want to have representation by women. So when the Dean's office called me up once I said, "Look, you're only asking me because I'm a woman, so give me a break and let me do my research."' Of course, in declining, one runs the risk of being considered a 'bad citizen', even though it is in a Republic of Science where she is often not accorded full citizenship.

TENURE

The contradiction between the tenure clock and the biological clock, for women, illustrates the taken for granted compatibility of the career structure of science with traditional assumptions of male youthful achievement. Despite the paucity of evidence that youth is associated with scientific achievement (Merton, 1973), the U.S. academic system is geared toward a forced march in the early years, allowing a slower pace later. This is exactly the opposite of the structure that would be preferable for most women. Until it is changed, there is little prospect of attracting a signficant number of women to careers at the highest levels of academic science.

The incompatibility of the seven-year race for tenure with the biological clock for child-bearing has obvious negative consequences for women's participation in high-powered academic science. A male faculty member told us that if women would wait until after age 35 to have children, there would be no problem. They would be able to pursue tenure single mindedly without interference from other obligations. He recognized that most women were unwilling to delay having children that long and thus saw no answer to this dilemma.

A graduate, now a professor at another university, reflected upon the relationship between the biological and tenure clocks. In discussing her plans for children she said: 'I take every day as it comes. It would be outrageously difficult. I would feel much more confidence if I had tenure but I would be 38 and I don't choose to have a child that late.' A faculty member's tenure review has caused an added measure of anxiety. She said: ' When it comes to the real facts that's when you feel discrimination. The pregnancy worries me. It's the wrong moment, always the wrong moment. It puts you on a slower track. Maybe they do see it like that. Maybe I've ruined my chances. They want you to sacrifice something. If the baby hadn't shown up, I would have pushed for an early decision. Now I will wait.' Career disruptions are often caused by the inability of the academic system to easily allow a modest reduction in workload. A supposedly temporary withdrawal is often the only option, but the expected return sometimes does not happen.

Departmental and university-wide efforts to make workplace child-care facilities more widely available would help. An infant care center in a neighboring school, discovered by one female graduate student, helped several women with children in one of the departments studied to carry on their graduate work virtually without interruption. The center, caring for children from ten weeks to three years of age, was an experimental site with a capacity of eight children. Although there were a few other facilities for older children affiliated with the university or located in the neighborhood, child care is still a major concern for parents. It has received more attention from companies than universities in recent years (NRC, 1993).

RESISTANCE TO CHANGE

If the objective is a significant increase in the number of women pursuing high-powered scientific careers, institutional accom-modations will have to be made for women who wish to combine family with career. Accommodation is currently made for faculty members, typically men, who found corporate firms or research centers; however, these time conflicts usually occur after tenure has been attained, whereas women's time conflicts involving family responsibilities tend to occur earlier in their career trajectory, prior to tenure, placing them at risk. While time conflicts at later career stages may affect colleagues' views of a department member, they seldom if ever damage the career.

To achieve equality it is not just a matter of opening up opportunities but of changing the structure of the academic system. Simply put, women are more vulnerable than men prior to tenure. Accommodation for time conflicts must be made for women faculty members with children. Women who wish to pursue traditional female roles along with a scientific career must be accommodated by allowing a longer time span before the tenure decision. This accommodation had been promised to one faculty member in our sample but subsequently was not allowed.

Even under the best of circumstances the academic structure is resistant to accommodating family needs. A female faculty member in another department was able to arrange a modest reduction in official time commitment involving a reduced teaching load. She reported that, in her department, 'The faculty have been very supportive of me having children. After my review I've had people say, "How can you do that and have children too?"' This professor adopted the strategy of reducing her work load and lengthening the time period before the tenure decision.

She said that, 'The university policy allows you to work part-time to have children . . . that part-time work stops the tenure clock for the percentage of time you are not working. Because of tenure, I didn't want to cut my [research] back by 50%, so I made an arrangement to work 70% and cut the teaching load. Everybody assumed, including the chair that this time off would not count for tenure. A year before I was supposed to come up for tenure the chairman brought it up to the provost because [it was found that] the clock was still running. If it had stopped, I should have had an extra year before I was up for tenure so I would have more time to publish and get my research done. I decided not to fight it because I was concerned how going through a fight would affect the tenure decision. I was quite worried when the case went before the engineering school who are all older men who were all looking at me not having worked full time.' In this instance, the outcome was favorable but the anxiety level, normally high about tenure prospects, had been raised even further by the difficulties that the academic structure had in recognizing the presence of children in her life.

A few years later she was involved in an effort in the Senate of her university to make reduction in work load for women with children an official option. Some of the participants in the debate suggested that it should be among a list of limited choices in fringe benefits, or that it should be equally available to men and that therefore it was too costly to be made available at all. This suggests that the academic system is still resistant to accommodating women's needs. This is not a call for a 'mommy track', with different and lower expectations of achievement and rewards, but a serious effort to accommodate the significant number of women who are not willing to forgo family and children prior to tenure. It is unrealistic to expect significant numbers of women to follow the male model. If the goal is to substantially increase the participation of women in high-level academic science, a female model will have to be legitimated. Acceptance of an alternative career model is crucial both to placing more women in faculty slots in the immediate short term and to providing relevant role models for a broader range of female graduate students.

Efforts at reforming the academic structure by reducing the 'time bind' for women are fraught with danger as arguments in behalf of change are often turned into negative reflections on women's scientific abilities. Even in the absence of accommodation to their needs, the relatively few women in the system have maintained their productivity (Zuckerman and Cole, 1991). One female professor has spoken up in faculty meetings in favor of extending the time before tenure review for women with children. She sees this recommendation as a double-edged sword, however, as pressing for reducing the demands made on women with children might jeopardize their status by supporting the notion that women with children cannot be productive. Of course, the extension could be made gender-neutral, with the same provisions offered to men with extensive responsibilities for child-rearing. Nevertheless, in practice, this would likely be seen as a measure to accommodate women. Without structural reform, barriers to entry and achievement will deter all but a highly persistent few.

TENURE STRESS

The definitive goal for all junior faculty members is the attainment of tenure. Its real and representational meanings cannot be minimized, so, it is relentlessly anxiety-provoking. The subtle and not so subtle differences around professional acceptance of women, as well as the unique stressors with which they cope, arouse and heighten anxiety for women, leaving them wary and overly self-observing.

As pioneers in departments which previously had no female faculty, for many there are no signposts by which to get a bearing. In an unreliable environment, a chemist describes how she looks 'and sees that there are no women around and I know tenure doesn't come easy. The faculty are older. Perhaps the younger ones see things differently. I can't ever know for sure what's in the back of their minds.' A young engineer is careful what she says, noting, 'They've never tenured a woman in this department. So it is tricky approaching senior people and not have them get offended by what you say when you're waiting for your tenure decision.'

Between colleagues, the strain and fear around tenure decisions manifests itself in one other significant way. To protect themselves from feelings of disappointment, loss and possibly anger directed at the department if a junior woman is denied tenure, some tenured women who wish to reach out and befriend junior faculty self-protectively avoid forming such relationships. Two tenured colleagues in the same chemistry department, now closest of friends, describe avoiding any social interactions before they received tenure. One admits that she 'didn't want to take the risk of seeing her go if she didn't get tenure.' Considering the detrimental affects of isolation for those solitary women without female peers, tenure is experienced as so uncertain that it can impede much-needed relationships and connections where they are possible.

It is near or at the point of tenure that some of the most disturbing indications of different treatment of women by the power structure occur. These range from remarks by a chair implying that a woman physicist should not worry whether she was granted tenure or not because 'you have two salaries. You have a husband' to an emerging picture that some young professors suffer reprisals for being too proactive on behalf of women students, resulting in denial or postponement of tenure. After learning that the chair had postponed her tenure decision, and had laid out a plan for the coming year that 'sounded like a thirty-year career plan', this activist faculty member finds herself wondering 'if he resented that I was serving as a faculty advisor. Did he resent my role as an advocate for the students? Did that have to do with being a woman, or a bit of a rabble-rouser?'

In a highly competitive academic environment gender differences in faculty experience are often ascribed to the normal workings of the system, even by some of the women who are discriminated against. There may be a simple lack of awareness of unequal treatment when some parameters are equal. For example, a committee representing senior female faculty members at the Massachusetts Institute of Technology recently found, to their initial shock, that they had lower salaries and smaller offices than their male counterparts.

When confronted with these findings, MIT President, Charles Vest, broke with long tradition of academic denial of gender discrimination. He pledged the full weight of his office to redress these specific grievances and undertook a commitment to broader change. However, more than a decade ago, a committee of graduate women in the electrical engineering and computer science department at MIT produced a report detailing their similar experiences of gender discrimination (Goldberg, 1999; Spertus, 1991; Female Graduate Students and Research Staff, 1983).

11 Differences between women in science

Overlying the differences between the male and female scientific worlds is another split, one within the female realm, that mirrors and refracts the larger gender division. Women scientists' perception of the obstacles in their path, and their response to them, create two dichotomous camps. One group is predominantly made up of an older generation of resilient women who stress a highly competitive, individualistic style that mirrors the traditional male stereotype. In contrast, younger up-and-coming junior and newly tenured women faculty members emphasize a more relational, collaborative approach within their research groups. We call the first group of women 'instrumentals', and the second 'relationals', reflecting their respective work styles. Notwithstanding such important differences, women faculty members who have thrived appear to have in common two significant characteristics. Firstly, they all identify sufficiently positive relationships with their own graduate school advisors as crucial to their past and present level of self-confidence, perseverance, and success. Secondly, although each is influenced by their own perception of a scientific style, all of these dedicated women labor to interpret an appropriate role as advisor to their female students.

Inescapable tensions exist for all successful women scientists, regardless of personal philosophy, around the role of advisor. The perceptions, attitudes and values which comprise a 'style' of advising and doing science are frequently a product of, or a response to, an earlier powerful relationship with one's mentor. The attitudes and values of the mentor-advisor are internalized by the apprentice and become the core structure by which an individual comes to form, and perhaps later modify, their own identity as a scientist. The tension between the relational and instrumental styles of women faculty reflects not only a generational shift among women scientists, but also the changing values of a new generation of male scientists. Influenced by attitudinal changes on gender issues experienced in their own personal lives, their relationships with women, and perhaps the feminist movement, some of these (usually) younger men have taken up a relational style. The differences among female academic scientists emerge most clearly in their role as advisors to students.

THE CONTRADICTIONS OF BEING AN ADVISOR

There is an increasing recognition that serving as a role model is complex and requires more than just 'being there' as a physical presence. Especially for younger tenured and non-tenured female faculty no issue presents more conflict or is filled with more angst than that of the role of advisor. How one advises, and particularly how one advises women graduate students, can become the very locus of self-definition. Not only can it bring forth the difficult issues and related feelings of being a woman in science, but it can produce a sense of responsibility for the next generation. Thus the needs that were or were not adequately filled as a student, or now as a faculty member – issues around birth and child care, balancing work and an outside life – will all be evoked through this role.

For men who have not been directly affected by these difficult issues, the role of advisor is not laden with these subjective meanings nor is it as emotionally charged. Women faculty tend to be deeply affected by their impact on their students, particularly on the women. This emphasis on relatedness and wish of many women faculty to nurture their students is more than simply a consequence of cultural 'socialization'. It may also be part of the dynamic of female adult development and the importance placed upon personal attachments and connections that transcend utilitarian motivations (Chodorow, 1978). At the same time, there is also a counter-pull on personal resources deriving from their own need for professional survival. As competitive players in a competitive business, women faculty members find a tension often builds up between their needs and the needs of others, particularly when they strongly identify with the issues faced by their female students. For the majority, it is a role not taken lightly nor is it without internal conflict.

The role of advisor is a complex task, requiring a great deal of emotional and intellectual energy. Women advisors constantly struggle with a varied range of issues, including finding a balance between acting for their students and becoming viable professionals themselves. On one hand, female faculty members must negotiate a competitive and judgmental environment and cannot afford to make themselves vulnerable. On the other, there is confusion as to what kind of relationship to establish with female students and what role to play in both their scientific and emotional lives. As one scientist summed up the dilemma, 'I've been asked to do a lot. To talk to undergraduates, graduate students, women's groups, all sorts of things. But if I'm not here in three years, then I'm not going to do anybody any good. So I can only do so much because I'm more worried about my retention. I try to do what I can. I think one of the reasons I wanted to go into academia was to be able to have an impact in one way or another on women's issues. I really want to, but I really can't do much better.' Accepting realistic limits on the time they can devote to women's issues is an especially hard lesson for female scientists committed to advancing the cause of women in science.

Current funding pressures increase the tension between the wish to mentor and provide support, and the need to remain productive and competitive. The competitive push sometimes makes it 'easy for me to forget that I had support.' However, inner conflict frequently arises when one knows the kind of advisor one wishes to become, but cannot afford to be, particularly around attempts to meet the child-care needs of graduate women. Although sensitive and responsive to the needs of one of her students, this biologist describes the inescapable inner turmoil provoked by her own needs: 'I'm trying to help her, but I feel like an ogre. It's a difficult one for me. I'm trying to be as supportive as I can, but then the work's not getting done fast enough.' The conflict between the demands of scientific productivity and family life requires adjustments in the organization of science in the university that transcend the advisor–advisee relationship.

Some advisors feel caught between wanting to meet legitimate needs and wishing for students to be assertive and competitive in order to succeed. 'Faculty have an ambivalent relationship with their graduate students. They want them to be one way for their own needs, as well as they want them to be a certain way because that would be good for [the student's] career.' Some faculty members guard against repeating the destructive characteristics of their own advisor since many are aware that 'you do as was done to you . . . sometimes the way I treat my group I'm shocked because I sound exactly like my advisor.'

Moreover, some women despair of being able to play any significant role for women students because they feel helpless to change the academic structure to make it more amenable to combining work and family. As one female academic analyzed the emerging female graduate student perspective on future faculty life, 'they see [my colleague] under all this incredible stress. She's trying to have a baby and these guys [male faculty members] give her a terrible time. So even when students get along with their advisor, when they look at her life, they don't like what they see.' The difficulties that female graduate students see their female professors encounter make them think twice, if they do not deter them entirely from undertaking an academic career at a high-powered research university.

Many faculty worry about how candid they should be with female students about their own difficulties in science. Concerned that they will frighten them, many say nothing at all. At the core there is an aura of helplessness as to how to not be discouraging. For instance, this advisor felt 'grateful' that a younger colleague had her baby at 30 rather than at 40 like herself so that her students would not become disheartened. She said, 'The students look at me and think, I would do anything other than have a life like she has. I admit that I communicate my difficulties, and I don't know how positive that is for encouraging other women. I worry a lot about it.' Again, the issue for the advisor is to find an appropriate balance between realistically depicting the barriers that women in science encounter, so that female students will be prepared to deal with them, and avoiding turning them away from scientific careers.

Some female faculty members, who heretofore did not take women's issues into account, have come to question their effectiveness as role models in the past. This reconsideration of previous practice often leads to new attempts to advise differently. A 42-year-old tenured woman acknowledges, 'It never occurred to me [at that time] that there were differences in men and women. In retrospect, I can see a thousand of them. When I got my first job they questioned me closely about whether I was serious about science, was I going to have kids. At the time I left there and came here, I never thought about these issues. I was moderately successful and I believed that the best thing that I could do for women was to be a successful professor. A lot of women still believe that.' The notion that it is sufficient to do good science to pave the way for future generations of female scientists is increasingly questioned.

When women students drop out, even female scientists who previously stood apart from women's issues may begin to re-interpret the advisory role. More attuned to the negative experiences of their students, they become more self-aware: 'I've been successful and something of a loner, and somewhat independent. Now that I have students of my own I'm worried. It's not unusual for successful and independent women to start out thinking they don't need anybody. But then they see the young women they care about and realize they might need support.' This realization has led some female scientists who previously adhered to the 'male model' to rethink their position.

Lastly, without collegial interactions to help understand similar experiences, young faculty members are left to re-invent how best to manage, supervise, role model, mentor, and do research . . . even when a critical mass of female colleagues is present. Since younger academics highly value interpersonal effectiveness, when problems arise they can be experienced as personal failure. Many young

professors discuss the importance of relationships in their labs:
'Nobody trains you how to be an advisor. They don't teach you how to
teach either. I found human problems are the most difficult part of this
work. Science is easy in comparison.' In another context, at a Cold
Spring Harbor conference on the biotechnology industry, seemingly
completely apart from issues of women in science, a female scientist
noted how useful a course on lab management would have been as a
part of her graduate training. Long past that stage, she had to learn it on
her own.

Women's emphasis on the role they wish to play in behalf of younger
women cannot be underestimated. The empathy for students'
difficulties with self-esteem and self-confidence comes from their own
experiences as women in science, and mentoring around these issues
evokes painful feelings and creates vulnerability. The most profound
struggle is whether to permit the close connection wished for by both
student and advisor, or to sustain what is a psychologically protective
distance for the self. As described by a biologist:

> 'I struggle with the issue of how strongly I model for other women.
> It seems that enabling young women to express what they think
> and how they feel is an important goal. When I think, would I do
> that . . . that would be hard. I'm starting to cry because there's a
> part of me that really wants to maintain a certain amount of
> distance because I identify so completely with them that I almost
> become overwhelmed.

This inner debate about what it means to be a woman in science is
matched and modeled by an exterior debate between proponents of two
opposing world views about how to conduct oneself as a female
scientist.

THE INSTRUMENTAL STYLE
Many older women scientists who have conformed to the traditional
'instrumental' male model find themselves confused by their
relationships with a new generation of young women who seek change

in the social relations of science. While these older women faculty recognize how the support of their own advisors was crucial to their development, they become confused and sometimes frustrated by new demands that women students make of them. Their students wish them to engage with them on a personal as well as a scientific level and, most importantly, to advise them about how to combine the roles of science and family. Instrumentals are forced to defend old beliefs in the face of new demands, admitting, as did this woman physicist, 'I have behaved like a man. If I got one of these [women's] surveys, I would throw it away . . . I don't discuss women's issues. I don't have time to get involved. I'm not involved. I went to their meeting and the women wanted to talk about day care. I don't have time. It's enough to keep on going.' Thus, many older women find themselves in a conundrum in which the notion of support is highly valued, but the demands made upon them to attend to issues of importance to female students seem alien, uncomfortable, and unprofessional.

This older generation of women faculty typically received support and strategic assistance from male advisors who were intensely competitive and individualistic. These men exclusively focused on their science, and expected and rewarded the same in their students. Even to consider having a family before tenure would impede such super-human efforts. Child-bearing and child-rearing during the early stages of a scientific career were declared non-issues. As this physicist recalled, 'My advisor sat down with me and said "No babies during graduate school . . .".' The senior generation of female scientists never denied the reality of gender bias, but their only solution was to work harder and become a superstar in order to 'prove' it fallacious.

THE RELATIONAL STYLE

As a new cohort of female faculty has entered the scientific arena, they have brought with them a more collaborative experience with their relatively younger male advisors. These men are frequently married to professional women, often with young families of their own. Not only are effective interpersonal relationships viewed by these younger men as an important strategic component to production within their labs, but issues of family, childbirth, and sensitivity to gender bias are treated as valid and often informally incorporated into the mentor relationship as relevant for strategic planning. As women assume junior faculty positions, they must evaluate how they will preserve and transmit these values while safeguarding their tenuous status within the department.

Whether tenured or not, women exemplifying the 'relational' approach emphasize collaboration and community within their own groups. Relationships among the members of the lab are important to research strategy, as faculty members treat each student as an individual with different needs and strengths. As an electrical engineering professor described her laboratory:

> We are all creating and taking and sharing responsibilities and experiments so we can interact together and contribute the expertise of each student so that they can feel like they are valuable . . . when I add people to the group, even if they are a 6.0 student [that is, off the measurement scales], if the chemistry is not right I will not add them in. We do things together . . . that make us know each other on a social scale. Then to reinforce the group activity in the lab, we have a buddy system . . . we will rotate the buddies so that everyone is working together . . .

The social organization in such a lab is lateral, in contrast to the traditional hierarchical model of the faculty member operating through a 'lieutenant', typically a post-doctoral fellow, in supervising each graduate student's work.

In contrast to the more singularly focused, instrumental women faculty, these younger women empathize with their female graduate students around issues of pregnancy and child-rearing, sometimes sharing the same dilemmas and even looking to their students for insight. For many senior women, however, obstacles such as these only exist subjectively; if you buy into them, you view yourself as a victim and cloud your scientific focus with extraneous concerns.

To the younger generation, prejudice and obstacles are not only perceived as real but are further exacerbated by conflicts over the compatibility of science with family and other roles. These contrasting modes of being a woman in science frequently give confusing signals to young women who seek role models whom they can emulate and who will provide them with the necessary 'truths' in order to succeed.

SUMMARY

In our analysis of women's graduate school and faculty experiences, we have focused on women's exclusion and disappointment within the structures of organizations where women are recruited, trained, and evaluated (Baron and Bielby, 1980). Our arguments focused on how the organization of science, particularly at the department and university level, differentially *treats and disadvantages* women and illustrated how women with human capital and career aspirations equal to or greater than their male peers are disadvantaged in their graduate school careers. We also noted how the women's conflict over balancing their family and professional lives creates unique needs because societal norms and routine practices work against women's careers even as men and society benefit from the traditional structures.

It is through this experience that the number of girls who pursue science decreases as children progress through the school system, while the loss of women increases. We may ultimately come to view women's withdrawal from science as a wholesomely adaptive response to marginality, based on such destructive experiences. We may also come to understand better the mechanisms used by an older generation of women in order to persevere, as well as appreciate a new generation of scientists who have begun to re-interpret science as a genderless and collaborative endeavor.

12 Social capital and faculty network relationships

If a man can write a better book,
preach a better sermon,
or make a better mousetrap than his neighbor,
though he builds his house in the woods
the world will make a beaten path to his door.

Ralph Waldo Emerson (1855)

In Chapter 6, we argued that differences in the network of contacts that a person has for exchanging resources and learning – their social capital – explain why and how women students are socially isolated and limited in their opportunities to form relationships, produce science, and gain support for their careers. In Chapter 8, we extended the argument to explain the experiences of newly minted women scientists who experience many of the same biases they did as graduate students, compounding the negative effects of their graduate school relationships and laying the groundwork for a perpetuation of the status quo. In the complex and political world of science, exclusion not only decreases a person's ability to acquire the knowledge needed to build a better mousetrap, but also decreases the ability of others to learn about and adopt it.[1]

In this chapter, we use original survey data on the social networks of men and women faculty members to further explicate our argument

---

[1] The 'Waldo' hypothesis is interesting in that it so squarely places human capital at the seat of innovation and success. Yet Emerson's own success appears to have been greatly shaped by his accumulation of social capital. Burt (1998) notes that Emerson was born into a poor family and through cultivated ties became noticed by Harvard's ministerial studies department. After graduation, he changed his name to Waldo, built his connections through lectures and the minister's association, and married the daughter of a well-to-do family whom he met at a speaking engagement. The question is would an equally qualified person who did not go to Harvard, have access to the minister's association, or marry well have enjoyed equal success?

and show that exclusion and access measurably affect not only perceptions of difference, but also career performance. Our research strategy triangulates our qualitative findings with survey data from a sample of male and female faculty members of science departments at an elite Midwestern university, a setting in which gender differences should be least likely to matter (Sonnert and Holton, 1994). By examining data on men and women faculty members who work side-by-side we can directly compare their experiences and the consequences of their experiences and explore the representativeness of conclusions.

We demonstrate that the cumulative effects of negative graduate school experiences and tenure-track positions that we described earlier shape attainment and work experiences, even among the select women scientists who have been able to cross the threshold from graduate student to untenured professor. We show that women experience more exclusionary and tokenistic practices in their collegial relationships than their male peers, that their networks are more isolated and alienating, and that these differences are associated with barriers to performance, even when no differences in human capital exist.

The chapter is organized as follows. First, building on the social capital concepts introduced previously, we review how social capital and network structure influence the careers of current female and male faculty members. Second, we describe the differences in the social capital of female and male faculty members. Third, we test the argument and show social capital's effect on the success of scientific careers in the 'publish or perish' world of contemporary science (Cole, 1992; Pfeffer, 1993).

SOCIAL CAPITAL AND DEPARTMENTAL RELATIONSHIPS
In previous chapters we argued that men and women faculty members experience separate scientific worlds. The male world is characterized by stronger social and professional ties than is the female world. Typically, men form close social ties with other male colleagues

within and beyond the department that facilitate access to collegial resources and information, which in turn help them to identify promising studies, manage labs, or learn the politics of tenure and publishing. In contrast, women typically report the obverse: their relationships with male colleagues tend to lack the close personal relationships and high levels of reciprocity they observe in male-to-male relationships within their departments. In this sense, the scientific worlds of women and men faculty who work side-by-side creates an ironic duality. Whereas male faculty work in closely knit social networks of exchange – in 'kula rings' – that belie the conception of the lone scientist finding truth in isolation from outside influences, this lone scientist conception is imposed on women, undermining rather than enhancing their ability to succeed.

In earlier chapters, our interviews with women faculty members discovered that men's and women's relationships varied along two dimensions, which we might broadly refer to as 'colleagueship' and 'reciprocation' – dimensions similar to those found in other professional environments (Podolny and Baron, 1997: 675–6). *Colleagueship* affected a person's sense of exclusion or isolation from their department and colleagues and was manifested in two ways. One indicator of colleagueship was the level of *social support* or friendship in the relationship. Did colleagues form friendships and exchange non-professional or private information in social settings that helped to manage the stress of publishing, teaching, and office politics? The other indicator related to how important the relationship was for conferring a positive *professional identity*. Did colleagues reflect expectations and sentiments that reinforced a positive image of one's ability to be a good scientist and to do good scientific work? Relationships high in colleagueship were reported to be important because they tended to reduce stress and provided a second opinion in a world where the quality and value of one's work is subjectively evaluated and commonly takes years to have a visible impact. The qualitative research suggested that men tend to have ties that are more socially supportive and confer more positive social identity than women, creating feelings of empowerment and inclusion in male faculty, and estrangement and resignation in women faculty.

Reciprocation affected a person's ability to access and exchange tangible professional resources. It also was manifested in two ways. One symptom related to the degree to which the relationship was characterized by an imbalance of exchange of resources. In particular, we found that women scientists were typically tokens in departments and often experienced 'token overload', meaning that women often had to shoulder tasks and responsibilities that their male peers did not. For example, did colleagues expect their papers to be proofed, exams to be proctored, or grants to be reviewed but not expect to do the same in return? Did women have to take on more committee work for the purposes of providing 'female' representation? The other symptom related to the degree to which the relationship was characterized by *power imbalances.* Did colleagues view the person as a professional equal or unequal? Did women have to work harder to prove their worth and gain the favor of colleagues than did male peers? In the next section, we reveal the results of our survey of male and female faculty members, with special attention to how the qualities of their department relationships differ.

GENDER DIFFERENCES IN QUALITY OF DEPARTMENT RELATIONSHIPS

To examine the differences in the relationships of male and female academics and their effect on attainment, we surveyed faculty members in six hard science departments (biology, biochemistry, chemistry, computer science, engineering, and physics) at a top-ranked, elite U.S. university. The questionnaire provided systematic data on the quality of the departments relationships along the dimensions of colleagueship and reciprocation. A social network was defined as the 'people you go to for vital professional advice or social support regarding work related activities.' Our constructs of Social Support, Professional Identity, Token Overload, and Power Imbalance were measured using the following items and a five-point Likert scale respectively.

(1) 'How important to your personal feelings of well-being is [contact's name]?'

(2) 'To what extent does this contact [name of contact] understand your particular needs as a faculty member?'

(3) 'If the following person [name of contact] asked you to do an activity that had no direct benefit to you (e.g. doing committee work, running an experiment, or proofing a manuscript), would you?'

(4) 'Do you agree or disagree with the following statement? I sometimes exaggerate the qualities (e.g. experience or expertise) of [contact] to show that I value their relationship.'

Other characteristics of the sample, design, and methods used in the survey are described in the appendix.

Consistent with our arguments and field data, we found that women report lower levels of colleagueship and reciprocation in the department relationships than do their male peers. Table 12.1 shows the means of our four measures of relationship quality. We present the relevant comparisons by three categories: all faculty, untenured

Table 12.1 *Quality of department relationships*

Department relationships	All faculty		Untenured faculty		Untenured women faculty	
	Men	Women	Men	Women	Critical mass	Tokens
Social support	3.7*	3.2	3.7*	3.2	3.4*	2.8
Professional identity	4.0	4.1	4.5*	4.2	4.2	4.0
Token overload	3.9*	4.3	3.9*	4.3	4.3	4.5
Power imbalance	3.7*	4.3	3.8*	4.4	4.7*	3.7

Note: An asterisk indicates significant differences in men's and women's responses by category and relationship type.

faculty, and untenured women in departments of critical mass and token status. Departments of critical mass and token status were defined in accordance with prior research (Kanter, 1977; Ibarra and Smith-Lovin, 1997). A department has a critical mass of women faculty if women compose more than 15% of faculty members. A department has a token status if women make up 15% or less of the faculty. (A department has a majority status if women make up more than 60% of the faculty.) Table A1 in the Appendix shows the exact percentage of women in each of the departments we sampled and indicates that biology and biochemistry have a critical mass, whereas chemistry, computer science, engineering, and physics have token status.

The results reveal that on average untenured men experienced their contacts as being more supportive of their psychological well-being than did untenured women. Similarly, untenured men scored their contacts as more enhancing of their professional identity than did women. These results suggest that within the department, male faculty members maintain relationships that provide more social support as well as conferring greater levels of professional identity than do women, a finding that reinforces the conclusions of our fieldwork. Furthermore, these differences suggest that men have more outlets for reducing stress and are empowered to achieve higher goals. Moreover, while these effects may be intangible their importance is hard to overstate because of the unique character of academic careers. The intense stress of tenure and the postponed feedback one receives about one's work means that these factors can both ease the psychological costs as well as maintaining aspiration levels of a serious scientific career.

The results also reveal that women report lower levels of reciprocation in their relationships. This finding is consistent with our interview data and suggests that the problems of colleagueship are compounded by relationships that provide few tangible resources. First, in terms of token overload, women report that they dedicate more resources to their contacts than are likely to be received in return. This finding is consistent with our interview data, which suggests that women faculty are expected to dedicate more resources to their relationships to compensate for the false perception that their intellectual contributions are lower than their male peers. Second, the data reveal that women are more likely than their male peers to have unequal power relationships with their colleagues. Women are statistically more likely than their male peers to experience the need to work relatively harder to prove their worth and gain the favor of colleagues.

Like the results regarding the level of colleagueship, these findings suggest that women's department relationships provide fewer of the tangible resources that are important for academic success. Their relationships required more time and resources to sustain than do those of their male peers, with apparently no offsetting level of resource inflows. Thus, they allot more time or effort to communicating and explaining their accomplishments and failures than do the men who are part of the demographic majority (Kanter, 1977). The implications of these findings are that women carry a double burden of meeting the requirements of academic success as well as the added provisions of token overload and unequal power relationships.

We also found that these problems were most acute for women in departments where they had token status (Kanter, 1977), and that a critical mass of women faculty members improved these problems but failed to reduce them to a level where male and female faculty members rated their ties similarly. This is an important extension of our findings and suggests that critical mass has positive effects on women's experiences in science, even though an increase in size may result in divided subgroups of women who are at odds with one another (Kanter, 1977). Indeed, this dividedness is partly suggested by the fact that critical mass improves women's perceptions of relationships but not up to the level of male faculty.

Finally, Table 12.1 shows the relevant comparisons by token versus critical mass status. Consistent with our argument, the data show that women in departments of critical mass are more likely to report relationships that have higher levels of social support and identity enhancement than are women in departments where they are a token minority. Similarly, their contacts provided more reciprocation than in departments where women were tokens, but were still less positive than men's. Conversely, the results suggest that women in departments where they have token status have a double disadvantage: they are less likely to receive either social support or important productive resources via their relationships. This suggests that while critical mass may be one factor that can overcome the barriers to success of women, other interventions are needed. We return to this in Chapters 13 and 14 where we compare different models of graduate education and their relative successes in reducing gender-related barriers.

GENDER DIFFERENCES IN DEPARTMENT NETWORKS

Our comparison of department relationships also contrasts the number of department ties that men and women faculty members report having for professional or social work-related advice. This analysis supplements our analysis of perceptual measures of relationships with a behavioral indicator – the assumption being that perceptions should correspond with actions (Marsden and Campbell, 1984; Ibarra and Andrews, 1993).

Our fieldwork suggested that faculty who named an intermediate number of contacts tended to have relationships with these contacts that were characterized by high levels of colleagueship and reciprocation. At first blush, this view seems counter-intuitive because more ties are normally assumed to provide better connectedness and access to information and resources. Our research and that of others on scientific careers suggests, however, that although a large network of ties *beyond* a department may be uniformly beneficial, a network of ties within the department that is too small or large may signal problems of integration into the department (Powell *et al.*, 1998). This is because an intermediate number of department ties strikes a requisite balance between having a set of collaborative relationships on the one hand and the ability to sustain meaningful relationships on the other hand.

As we argued in earlier chapters, a network of department contacts that is few in number lacks a wide enough base of social and professional resources to be effective even if the ties within the network are close, giving, and supportive. These kinds of networks are too insular. Our fieldwork suggests that women are more typically in these kinds of department networks than are men. Conversely, a network of strong department contacts that is too large requires a large investment in time and energy to maintain – draining important resources away from other productive activities (Uzzi, 1998). For instance, these processes were exemplified by a woman professor who commented on the compromise between forming many close attachments within department and the need to spend time and resources on other professional activities or building ties to people in other departments. She said, 'The bigger concern is whether energies I spend here are energies I don't spend somewhere else. And somewhere else being things that affect the nurturing outside [relationships].'

Table 12.2 displays the number of strong department contacts named by our respondents. Strong ties were measured by asking respondents, 'What faculty in the department would you feel comfortable approaching for important personal or professional advice?' Table 12.2 suggests that men and women untenured faculty members have similar distributions of strong ties in their departments. Consistent with theory, men tend to have an intermediate number of strong ties – the modal number is three. Somewhat surprisingly, the women in our sample also reported having an intermediate number of strong department ties. However, when these results are disaggregated by critical mass and token status, they suggest that a critical mass of women promotes the development of a well balanced network of strong ties, a finding consistent with our expectations about the role of critical mass. Although the actual numbers make unambiguous inferences troublesome, they suggest, in line with our fieldwork, that in departments with critical mass, women's strong tie networks look like those of their male counterparts. They have neither too few nor too many strong ties within the department. In contrast, women in

Table 12.2 *Number of strong department contacts named*

Ties in the dept. Number	All faculty				Untenured faculty				Untenured women faculty			
	Men		Women		Men		Women		Critical mass		Tokens	
	Freq.	%	Freq.	%	Freq.	%	Freq.	%	Freq.	%	Freq.	%
1	11	13	0	0	4	15	0	0	0	0	0	0
2	10	11	3	23	0	0	3	27	3	43	2	50
3	39	46	3	23	16	59	3	27	1	14	0	0
4	9	10	0	0	2	7	0	0	0	0	0	0
5	15	17	7	53	5	19	5	45	3	43	2	50
Total	84	100	13	100	27	100	11	100	7	100	4	100

departments of token status tend to have either too few or too many strong ties – suggesting that their networks may lack the capacity for collaboration or have high maintenance and opportunity costs. Thus, the patterns of results from our interviews, the quantitative data on the perception of the quality of department ties, and the quantitative data on the number of close department ties converge on a similar theme. Women in departments of token status, as opposed to women in departments of critical mass, lack the strong tie networks that promote empowerment and opportunities for academic success within the department.

GENDER DIFFERENCES IN INTERDEPARTMENTAL NETWORKS

Our fieldwork also argues that interdepartmental ties play an important role in the careers of women faculty. While departmental ties are important for collegueship and the reciprocal exchange of information and resources, interdepartmental ties are important for subfield visibility as well as access to novel or specialized information, needed for research breakthroughs but not represented in the backgrounds of local faculty specialists. Consequently, inter- departmental ties are critical for building reputations and accessing expertise that is unrepresented at the department level. Our fieldwork suggests that women's networks of ties beyond the department were significantly smaller and less diverse than men's faculty – a consequence of the accumulated disadvantages experienced in graduate school and their faculty experiences after graduation – and a barrier to their attainment.

In Chapter 2, we illustrated the importance of interdepartmental ties for gaining information when we described how they promoted Watson and Crick's discovery of the microstructure of DNA. In contrast, the absence of interdepartmental ties inhibited a similar accomplishment by Rosalind Franklin despite the fact that her laboratory had produced the first photos of DNA's structure. Specifically, Watson and Crick's interdepartmental ties to chemists in other laboratories (in particular the chemist who found the mistake in the chemistry text) opened up new opportunities for discovery by providing channels by which to import exciting new ideas and methods from other fields. And while the discovery of DNA's microstructure is an extraordinary case, the same principle holds in many ordinary cases that are arguably as consequential because of their prevalence. For example, a female graduate student recounted her experience of forming a tie with someone outside the department and described how it became a unique and valuable source of professional knowledge and advice.

> [I] . . . went to a grant writing workshop and saw a woman [from another department] there who actually stood up and asked questions. I actually went up and asked her, how do you do this? I'm fine on a one to one, but not in a big group . . . She said, "I sit in the front of the room. Then . . . It's just me and the other person talking." What a great idea. Later on I called her and I told her I was feeling very isolated and I'm wondering if you're feeling the same

way. She took the time to help. That makes me more likely to help someone else. I felt very touched by it and very lucky.

Our arguments explain the benefits of interdepartmental ties by building on and extending research on the social networks of professionals and entrepreneurs. Consistent with our fieldwork, these approaches have found that the social networks of professionals are made up of two types of connections that can be called *strong ties* (i.e., intradepartmental) and *bridging ties* (i.e., interdepartmental).[2] As noted above, strong ties are characterized by frequent interaction and usually involve collaboration (e.g., reading papers, doing conferences together, sharing committee assignments, co-authoring) or sensitive information (e.g., hidden dress codes, department politics, secrets). Unlike strong ties that require substantial resources to maintain, bridging ties require relatively infrequent contact and usually involve professional acquaintances. The key characteristic of bridging ties is that they are more likely to exist between persons in different professional circles, thereby becoming bridges over which new ideas flow between two otherwise disconnected research teams that could benefit from one another's knowledge. Moreover, bridging ties access information widely because they are likely to link local department networks together, and thus provide crucial viaducts to resources in

---

[2] A fuller treatment of these theoretical concepts can be found in Granovetter (1973), Burt (1992), Bian (1997), Podolny and Baron (1997), and Uzzi (1997; 1999). Readers from the hard sciences should see D.J. Watts and S. H. Strogatz's 1998 article in *Nature*, which explores the mathematics of networks in biological oscillators, Josephson junction arrays, excitable media, neural and genetic control networks, spatial games, and other self-organizing systems. Their insight is that connections between persons are ordinarily assumed to be either completely regular or completely random. But many types of networks lie between these extremes. On the basis of this assumption, the authors explore simple models of networks that can be attuned to this middle ground: regular networks 'rewired' to introduce increasing amounts of bridging ties. They call these reworked systems of local and bridging ties 'small-world networks', by analogy with the small-world phenomenon popularly known as six degrees of separation. The small-world phenomenon suggests that two strangers anywhere on the planet, such as a falafel maker on a street corner in Jerusalem and Marlon Brando, can be connected by a minimum of just six intermediaries (e.g., a friend of a friend and so on). This notion underscores the social network principle that the right contact can help individuals gain access to resources and information circulating in worlds very distant from their own.

fields like science where knowledge is fragmented and dispersed among many persons.

Our fieldwork suggested that bridging ties position a researcher to learn of new breakthroughs, to get important papers before they are published, to learn where competing researchers are investing their resources, or to import techniques from other disciplines into their own. A large network also generates channels for presenting work, educating users about its importance, and disseminating ideas prior to formal review. Finally, because bridging ties, unlike strong ties, require only modest amounts of resources to maintain, the larger the number the higher the faculty member's social capital. Faculty members who have only strong local ties to members of the department are likely to be cut off from these important channels of information and knowledge because their local department contacts share comparable viewpoints and resources.

A women professor approaching her tenure review revealed a sharp and vivid example of the benefits of a rich network of bridging ties. She commented on the importance of having a social network composed of both strong local ties as well as numerous bridging ties to faculty in other departments who would referee her case.

> So there are two issues, one what the inside letters would look like and what the paper case would look like; that is, the vita. But then there is also what the outside letters are going to look like . . . The way that I am supposed to decide officially from whom to solicit letters is by who knows my work well. I actually have a friend who came up for promotion who picked the seven most important people in his field. Needless to say he didn't get promoted. You just don't do that. You pick people to whom you have been sending your publications; you pick people who really know your work. My sense of how you do this is you have a sense from these people whether they are going to write you good letters. That is, at least you know that they think your work is good. You don't have to walk up to someone and say what kind of a letter will you write me, you know what that person thinks.

PATTERNS OF MEN'S AND WOMEN'S
INTERDEPARTMENTAL TIES

Table 12.3 shows the distributions of the number of bridging contacts outside the department. Bridging ties were measured by asking respondents to name the individuals outside the department that they could call if seeking professional information. In line with the previous section we break down the analysis by gender, untenured, and critical mass categories. Consistent with our fieldwork and qualitative interviews, we find that women cite fewer bridging contacts than men do for both the full faculty category and untenured faculty category. Women in departments of token status also report fewer contacts than women in departments with critical mass. These findings show that women, especially in departments of token status, have fewer bridging ties than men, suggesting that they are weak on a key dimension of professional success.

Taken together, a comparison of male and female faculty network within the department and outside the department converge on a

Table 12.3 *Bridging ties: Number of faculty contacts named outside the department*

Number of bridging ties	All faculty		Untenured faculty		Untenured women faculty	
	Men	Women	Men	Women	Critical mass	Tokens
Mean	9.2*	6.3	4.7	4.4	7.0*	4.7
Standard deviation of mean	15.7	6.4	4.9	3.2	7.7	1.2
Minimum	0.0	0.0	0.0	0.0	0.0	3.0
Maximum	99.0	25.0	20.0	10.0	25.0	6.0

Note: An asterisk indicates significant differences in men's and women's responses.

similar and crucial finding: women's networks tend to be poorer in social capital than those of their male peers. While this finding is expected given the fieldwork, the quantitative analysis reveals the sources and patterns of difference along gender lines. Moreover, it demonstrates that the self-reports of women faculty members capture their reality as well as their view of the reality of their male peers. Finally, the triangulation of results reinforces our fundamental conclusions about the sources of women's disadvantage in the hard sciences, and establishes pointers to underlying biases and strategies for overcoming them. In the next section, we examine the consequences of social capital.

ARE GENDER DIFFERENCES IN SOCIAL CAPITAL EXPLAINED BY DIFFERENCES IN OUR HUMAN CAPITAL?

Table 12.4 presents the mean differences between women and men faculty members along our three analytical categories: (a) all faculty by gender, (b) untenured faculty by gender, and (c) untenured women faculty by critical mass of women faculty in the department. Table A2 in the Appendix shows that men and women differ little in their human capital, a finding consistent with prior research on gender stratification in the professions (Pfeffer and Ross, 1982; Davis-Blake and Uzzi, 1993; Schneer and Reitman, 1993; Stroh and Brett, 1996; Uzzi and Barsness, 1998). Age and education level of the sample does not differ significantly when broken down by gender, tenure, or critical mass. The ages of the men and women show no significant differences, except for the difference between untenured women in departments with a critical mass and token status. Our three measures of work experience – years since the Ph.D. (professional age), number of years employed, and number of years in a post-doctoral fellowship position – show no significant differences. Both untenured men and women faculty members have had approximately five to six years of work experience and have spent two or three years as a post-doc before starting their present tenure-track position. Similarly, this sample of untenured women and men are approximately one to two years from their tenure decision which suggests that their understanding of their department is comparable and unlikely to change significantly before their tenure decision.

Several comparisons of the backgrounds of the women and men were also conducted. As expected, the education levels show no statistical differences by gender. The schools where faculty members received their doctorates are of similar status, and their Ph.D.s are usually from an institution within the top-ranked 15 universities. This is most likely a consequence of the fact that the hiring institution is a prestigious school. Married male faculty members are more likely to have children than are their female faculty peers (no single or divorced faculty members reported having children). Eighty-three percent of the untenured men versus 55 percent of the untenured women ($p < 0.8$) have children. The average age of the untenured women in this sample (41 years) suggests that these women have chosen not to have families, if they do not already have them. One reason for this effect may be that the sample represents 'survivors' (i.e., all women who wanted to start a family have left the sample). Another interpretation is that this sample

Table 12.4 *Respondent's human capital characteristics*

Variables	All faculty		Untenured faculty		Untenured women faculty	
	Men	Women	Men	Women	Tokens	Critical mass
Age	49	46	45	42	36*	45
Professional age	21	18	16	15	9	18
Years employed	14	10	6	6	7	5
Years in post-doc	3	2	3	2	3	2

Note: An asterisk indicates significant differences in men's and women's responses by category and characteristic.

of women have self-selected themselves into positions predicated on the assumption that they would forgo raising a family.

Finally, 83% of the untenured women in departments with critical mass ($p < 0.005$) have children versus zero percent among untenured women in departments without a critical mass. In sum, there are no significant differences in human capital among members of the same sample. They are all talented and well educated, and share similar work histories. However, there is an important difference between men and women in that a significant larger number of women than men participate in the tenure race without having children.

SOCIAL CAPITAL AND RESEARCH PRODUCTIVITY

The final question we explore is whether differences in social capital translate into differences in research productivity, holding constant human capital. We review the results of a regression analysis that linked social capital differences to research productivity using the popular convention of number of publications. While this measure does not account for conditions of work and job fulfillment, it is the dominant measure used to study career attainment and plays a disproportionately large role in promotion decisions (Zuckerman and Merton, 1972; Seashore *et al.*, 1989; Cole, 1992; Pfeffer, 1993).

In organizing our analysis, we look at three aspects of women's and men's networks that follow from our above examination of the quality of collegial relationships, number of strong ties, and number of bridging ties. First, we examine the predictive effects of *token overload* and *power imbalance* on publication rates, two empirical indicators of our underlying relationship constructs of collegiality and reciprocity respectively. We expect both of these factors to decrease a faculty member's publication success because both represent a lack of reciprocity in exchanges of tangible resources that are crucial for producing research. Following previous research, we do not hypothesize effects for social support or identity enhancement because these factors have been found to affect job turnover and job fulfillment rather than productivity per se (Podolny and Baron, 1997). (In analyses not shown, these variables did not have a statistically significant association with productivity.) Second, we examine the effects of the number of *strong* ties in the department. We expect faculty members who have insular or expansive department ties to have less attainment than those who keep an intermediate number of department ties. Our argument is that faculty members who keep too few ties are marginalized socially and/or politically in the department. Those who maintain large numbers of contacts experience alienation because their relationships tend to be shallow and low in trust. Third, we examine the effects of bridging ties and expect that bridging ties will be positively associated with publication success. We also include in our analysis standard control variables for human capital and department characteristics to isolate the net effects of social capital on research performance. The Appendix provides a full description of our statistical model and measures.

Table 12.5 reports the results of our regression of research productivity on social and human capital. Model 1 is the control model and contains variables that measure an individual's human capital and control for the statistical properties of their networks. It is presented to demonstrate that the predicted variables hold net of controls. Model 2 contains all the network, human capital, and statistical control variables. A positive and significant coefficient for an independent variable in these models indicates that it is positively associated with the level of research output.

Consistent with our expectations, we found that relationships high on power imbalance tend to reduce academic productivity. We also found that too few or too many strong ties negatively affect attainment, while an intermediate number of department contacts is positively associated with attainment. This suggests that either an under-investment in strong ties or an over-investment in strong ties hurts research productivity. Having too few strong ties decreases possibilities for collaboration and support, while too many strong ties have maintenance costs that override their benefits. Finally, we found that measures of both weak ties (number of contacts beyond the department) and number of co-authors are positively associated with

Table 12.5 *Ordered logit regression of social and human capital on research productivity*

	Model 1	Model 2
Social capital		
Token overload		–0.2542
		(0.3412)
Power imbalance		–0.5258*
		(0.270)
No. of strong ties		30.140***
		(10.080)
No. of strong ties squared		–0.615***
		(0.1712)
No. of bridge ties		0.0880**
		(0.0420)
No. of co-authors		0.0696***
		(0.0206)
Human Capital		
Gender (1=Male)	1.05*	0.0939
	(0.616)	(0.6938)
Tenure (1=Yes)	10.54***	10.062*
	(0.497)	(0.5726)
Professional age	0.036	0.1197*
	(0.056)	(0.0690)
Age	–0.009	–0.0899
	(0.054)	(0.0728)
Controls		
Network turnover	–0.019	0.1588
	(0.149)	(0.1673)
Average duration	–0.006**	–0.0068
	(0.003)	(0.0042)
Research budget	0.001	0.0008
	(0.000)	(0.0012)

Table 12.5 *(cont.)*

	Model 1	Model 2
Post-doc (1=Yes)	0.128	–0.0898
	(0.462)	(0.5362)
Model log likelihood	–930.452	–680.668
Pseudo R^2	0.1232	0.3557
Low productivity cut point	–20.28	–50.047
	(10.82)	(30.279)
Medium productivity cut point	–0.015	–20.004
	(10.76)	(30.259)
High productivity cut point	10.60	0.558
	(10.76)	(30.248)

$^*p < 0.1$; $^{**}p < 0.05$; $^{***}p < 0.01$. Standard errors are in parentheses. N=97.

research success. Consistent with our arguments, these results suggest that a large network of bridging ties aids timely access to intellectual capital.

These effects suggest that one of the underlying barriers to the success of women scientists is the structure of their social networks. Results of both structure and relations point in the same direction. Network structures composed of an intermediate level of strong department ties and a large network of bridging ties beyond the department are consistently associated with publishing by improving the ability of a researcher both to gain access to novel information that is circulating in other networks and to collaborate productively with close ties within the department on research projects.

SUMMARY

The role of kula rings outlined in earlier chapters is symbolic of the structure of relationships in contemporary science. Like kula rings that are founded on exclusivity and selective access to critical resources, relationships among faculty members follow the same principles of exchange and produce similar outcomes. In the popular parlance of science, the kula ring creates and allocates, through connections among scientists, the social capital that transforms their human capital into productive assets and conditions the experience of their work lives.

This chapter reinforces our conclusions about the social structural conditions that prevent women's full participation in scientific careers, even for those select women who attain faculty status despite having endured the barriers of gender socialization, overt discrimination, and conflicts between work and personal lives. A next step is to focus on the sources and consequences of social capital, and the strategies that can overcome its dark side and increase its benefits for women. Younger male faculty members express an understanding and interest in building more productive cross-gender and gender-inclusive networks, yet new strategies are needed and other problems exist. While women fare better in departments with a higher proportion of women, an increase in the number of women in a department sometimes, paradoxically, does not automatically produce positive effects when women split on key issues, some allying themselves with traditional male faculty members. Next, we examine what strategies have been used, what strategies furnish new possibilities, and what strategies are likely to overcome the problems we have identified.

13 Negative and positive departmental cultures

Ultimately departmental reform is the means to overcome the exclusion of accomplished women from full membership in the Republic of Science. In our most recent study, we were interested in identifying the characteristics of those graduate departments which showed the most and least improvement in the recruitment of women and conferring of the Ph.D., based on National Research Council (NRC) statistics from 1974 to 1990. In electrical engineering, the number was too low to generate meaningful data before 1977, and computer science had not been separated as a distinct discipline until 1978. In light of this, the time periods considered for these two disciplines were 1978–1990 and 1977–1990 respectively. What emerged was a range of departmental cultures. At that end of the spectrum where numbers of American women graduate students and/or degrees conferred were lowest, was what we call the 'Instrumental Department'. While most departments that we studied reflect the negative attitudes toward women in science, we also identified several 'Relational Departments' where positive cultural shifts are occurring.

THE INSTRUMENTAL DEPARTMENT

Not surprisingly, morale was lowest and isolation of women highest in instrumental departments. Many had no programs for women students and if they did, fear of stigma around joining was high. As a tenured woman preparing to leave for industry described the situation, 'How many faculty hires in the last 10 years? Zero. How many women interviewed? Zero. How many women students are supported? There was one several years ago. Maybe one now, or is it zero? The numbers are extremely grim.' It was not unusual for there to be only one woman faculty member present in the most depressed environments. As the lone woman professor, waiting for tenure for seven years in one such department, said, 'Who do I talk to? I feel lonely. I've always felt like that . . . I feel good seeing my picture in the front of the building, but there is only one [female].' Although numbers of women on faculty in all departments only rarely reached parity with males, instrumental departments had especially low levels of females or even none at all.

The severely instrumental department reflects a power structure which resides in the hands of a much older group of eminent male scientists who 'are resting on their Nobel Prizes . . . the imperial clan is always watching.' In the most hostile departments, generational attitudes were cited rather than that of gender. 'One of my biggest problems here is gender bias from the older faculty,' said one female professor. 'I never have worries like that from people of my own age.' Anxiety and feelings of powerlessness are very high and there is a sense of not 'ever knowing for sure what's in the back of their minds.' Nevertheless, the strong presence of instrumental attitudes among an older male generation of scientists and their relative lack among the younger generation augurs a significant change, favorable to female scientists.

Earlier studies of personality correlates and emotional characteristics of mature, distinguished scientists using projective tests (Cattell and Drevdahl, 1973; Eiduson, 1973; McClelland, 1973) shed some light on the experience of women faculty members in those least hospitable departments. Briefly, scientists were found to be more withdrawn than the general population, emotionally constricted, and controlling. As a group they avoided or were 'disturbed by complex human emotions' and were 'intensely' identified with that which was masculine. Lastly, the proving of oneself and the construction of identity through scientific work served to ward off sociability needs. Regardless of discipline, the overwhelming majority of departments in our study in some way reflected aspects of this model.

THE RELATIONAL DEPARTMENT

We have also identified a few 'Relational Departments' with a collegial and cooperative atmosphere that provide the safety to take the risks necessary for innovative work and the collaborations necessary for networking. Graduate and faculty women have created a grapevine to pass on information about departments with a relational culture that are often highly ranked but not at the very top of their field. Since some of the most talented researchers are women and men who want to follow a non-traditional path, departmental reform can be an academic mobility strategy, as well. A number of tenured women faculty who had struggled for recognition and status in prestigious graduate schools and post-doctoral programs that were highly competitive and hierarchal also reported accepting faculty appointments in more relational departments, hoping to repair the isolation and stress they had previously encountered. As advisors to graduate students, other successful women are directing their students to such departments as well.

Typically, departmental culture changes when an individual male, with a key role in the power structure, acquires feminist values. Such a personal transformation can translate into organizational change, especially when colleagues experience similar life changes. Women students and faculty members report being attracted to relational departments by interpersonal interactions during interviews, a sense of personal concern by the faculty committee, and an impression of 'happiness' and well-being among members of the department. Thus, relatedness, emotional closeness, 'the expectation of mutuality and the sharing of experience leading to a . . . sense of well being' (Kaplan and Surrey, 1984) are not only intrinsic criteria for choosing the department, but also describe elements of female developmental theory of the 'self-in-relation'. Most significantly, these milieus are relatively free of the enervating stress associated with the anxiety and defensive maneuvers required to be accepted and acceptable in highly instrumental departments.

In an instrumental department, interpersonal interactions are minimal and open communication avoided; the opposite condition holds in a relational department. Reflecting on her past experiences before joining a more women-inclusive faculty, a biologist felt that she 'had been in all-male environments long enough to know that it was important to minimize their anxiety. I became very good at only talking science and calming them down. It was energy consuming.' By contrast, a female faculty member in a relational department reported ' . . . a collective understanding. When I speak to [male faculty] I have the feeling I'm communicating with them as people. There is a recognition of the value to be had from cooperation. The emergence of individual empires is discouraged. A strong belief here is to preserve an environment which is as cooperative as is possible.' Following the model of relational research groups, there is an active commitment to sustain this milieu through the careful hiring of like-minded academic staff.

At a university in which both the chair of the department and an upper-echelon administrator were minority group members, women felt that their personal experience with discrimination encouraged empathic dialogue regarding women's problems. Moreover, this leadership diminished the power of the old boy network and supported affirmative action hires. 'Those people in their lives had really experienced discrimination and it changed how they behaved and made an enormous difference to us. We didn't always agree with [the chair], but we always felt that we were having the same conversation. This guy understood what we were talking about. That made a huge difference to this department because he never wanted to go out and hire honchos. He always wanted to hire young people and build them up. It is an open search.' The chair functions as a representative in relational departments, discussing issues with all faculty members before taking action.

Through personal contacts at scientific meetings and reports passed over e-mail lists, women are increasingly aware of 'good' and 'bad' departments and direct their applications accordingly. While the science being done in the department or by a faculty member often initiated a candidate's interest in the school, the emotional gratification of the interview process, together with a preference for a collegial research environment, influenced the candidate's final decision. Thus, selecting this particular department was a means of recapturing a significant professional and personal growth experience that had promoted self-confidence and emergence of a scientific self-identity. In the department mentioned above, a female academic model based on interpersonal relationships, affiliation, and nurturance had become accepted as legitimate and had even become the departmental norm. This was in strong contrast to another research site, where women's expression of a need for these characteristics in the laboratory environment was derided as a desire for dependence and emotionality by the adherents of the patriarchal system that was in place.

The context into which reforms are introduced is critical to their acceptance; the culture and organization of departments plays an important role in whether reforms will be accepted or rejected. A professsor who can mobilize a strong network on behalf of equality can transform a department. On one campus, the women all agreed that the ethos of a physics department had been changed through such an individual's efforts. As a female graduate student put it, 'My experience is that one person in an influential position can make a huge amount of difference.' This individual's goal had been to convey confidence and to help every student get through. According to a female graduate student, this individual was '... a tremendous influence on the whole tone of the department which made the place actually wonderful and people told me about this department. I never realized how much it was tied up with this one person.' On the other hand, grass-roots efforts can be undermined by 'old boys' if their power is entrenched and their numbers large. In another department, the chair 'created . . . [a positive] atmosphere here. He provided a strong presence. The conflict was with the people over 50: the old Guard. [The chair] finally resigned.' Strong networks can resist as well as assist change; those in power in a department can legitimize or delegitimize female affiliation.

ALTERNATIVE SCENARIOS

We identified several different scenarios of female faculty experience depending upon fit or lack of fit with departmental structure. These included female scientists who were attempting to follow (1) a relational style in an instrumental department, (2) an instrumental style in a relational department, and (3) an instrumental style in an instrumental department. There is a tripartite model of tension in which one of three scenarios is possible in any given department depending on its social construction. Each model describes an internal dilemma with which women faculty may struggle even when there is an apparent fit between an individual and their department.

Relational advisor/instrumental department

The tension in this situation is typically between a younger female faculty member who advocates on behalf of her female students and a senior woman faculty member who embraces the intensely competitive instrumental style. Sometimes relationships can be established far more easily with younger male faculty 'who share the same values.' Although she understands the history behind their behavior, this tenure track biologist describes her discomfort with older, pre-eminent women in her department who reflect the male model: 'I always end up not liking a lot of the women who make it to this level of science, enough to really not want to hang around with them. You just have to go through a lot of shit sometimes to really get there. Sometimes that means that you don't really care about anything else or anyone else. Sometimes that means that you face taking on the persona of a male scientist you don't think a whole lot of in terms of being aggressive and competitive.' Inexperienced, untenured women are not only alienated, but in several instances where their relational and proactive style has been conspicuous, tenure has been delayed or denied.

Instrumental advisor/relational department

For a female faculty member whose previous personal experience has been that she must prove herself to be exceptional and put aside all non-scientifiic interests, a department with a collaborative approach to mentoring induces a different inner turmoil. Such women view any detour from the individualistically competitive way of doing science as a disservice to their students. To recognize the personal issues that female students struggle with is experienced as a betrayal of what they believe is necessary to succeed. This lends itself to tension not only with their women students, but the department itself.

Professional pressures in tandem with forgetting early support systems of their own may also blur the picture, as reported by a veteran woman advisor. 'Sometimes it's easy for me to forget that I did have support. I think in retrospect you tend to forget the insecurities. The time you needed something extra.' Instrumental advisors, even in a relational department, may still be struggling with how to negotiate a competitive scientific environment and may not be able to afford to expose their own vulnerability.

Instrumental advisor/instrumental department

Women faculty members who identify with the traditional 'old boys' are perpexed about the tension that arises between them and female graduate students. Periodically such women find themselves with all-male laboratories and are at a loss to understand why female students gravitate to younger male advisors. Although this older advisor is cognizant of the self-doubts of her female students and recollects her own need for support, as lab director she becomes defensive about interpersonal demands she does not always understand. 'They think you are going to be very warm and supportive. I think I am a nice person. I certainly care a lot. But a professor has a certain duty to say "You have to get back on track or this is it . . ." If you're going to be in this field then your job is to criticize yourself every day and never get too down on yourself.'

However, as female graduate students have become more forceful in articulating their needs, women faculty members are forced to question their belief that women should work harder than men in order to prove their worth. Having ignored the women's movement of the early 1970s, owing to their exclusive scientific focus, senior women may find that the vicissitudes of young incoming faculty and students provide an 'eye-opening experience'. As a result, senior women faculty have had their consciousness raised by their women students in a number of such instances.

A heightened consciousness of discrimination is, of course, only the first step. Once individual attitudes change, the next step is to effect organizational change, which can be more difficult. The high powered academic science department with its competitive identity is typically highly resistant to change. Nevertheless, some improvement for women in science has taken place in recent years. In the next chapter we delineate several change-making strategies, their costs and benefits.

14 Initiatives for departmental change

This chapter provides descriptions of the classes of interventions – top down, bottom up, and idiosyncratic – that occur in academic departments attempting to bring about gender equality. We analyze the pros and cons of each of these types of intervention. Our proposal is to help administrators and policy analysts understand what kinds of interventions, given their limitations and advantages, bring about the best outcomes under different circumstances. Later in this chapter, we explain in detail how departments can use specific practices to change, develop, or enhance these interventions through task redesign, social networks, or university–industry relationships.

To study programs systematically, we delineated four groups of departments in each of the five target disciplines: biology, chemistry, physics, computer science and electrical engineering. The first group of departments had initiated programs whose stated objective was to be more inclusive of women. The other three groups were delineated on the basis of outcomes, as recorded by the National Research Council's (NRC) annual compilation of doctoral degrees granted. In each of the five fields for the decade-and-a-half up through 1990, we selected the ten departments that had graduated the highest proportion of female doctorates, the ten that had graduated the lowest, and the ten that showed the most improvement in women's graduation rate across that period. We then selected the two departments from each group that displayed the most consistent numbers and trends.

TYPES OF PROGRAMS FOR WOMEN IN SCIENCE
Programs are interventions from above and below that attempt to

repair the quality of women's educational experience, partially making up for significant deficits in the course of attaining the Ph.D. degree. Sometimes programs provide encouragement and advice, substituting for informal social venues in the department that exclude women. Other programs provide academic support, providing a parallel structure of study groups for women. Still others combine both characteristics, making available mentors to supplement gaps in the department's advisory system or counteract poor treatment of women by official academic advisors.

The basic root of all programs is the presence of a skilled individual to provide to women support, guidance and an independent perspective that compensates for the faculty's inability to do so. A discussion group leader described how

> the women talk about their problems and the thing that comes up most often is "How am I going to have an academic career and have children? How am I going to do both?". They're being taught by the males in the department about this standard of excellence and how they're going to go out and be the cream of the crop. Well, they're quite aware of the amount of work that their advisors do. The other female [faculty] member in the department is not married, she has no children, works constantly and the women are quite aware of that. Are they going to be required to give up any thought of a family life [when] they don't want to?

A female graduate student said

> Maybe it's particular to the sciences, but you get very closed in and you need someone to say, "I'm here" to make your path a little smoother. As a woman, if I knew there was somebody there I could talk to or would involve other students, that would change my perspective on the environment. I would feel that it was more of a community and more supportive. Even if I never took advantage of that service, I would know it was there.

A scheme to foster female graduate students' personal and professional development and an ethos of inclusion from an official departmental position is the next step.

The failure of many departments to adequately mentor women, whether arising from an unwillingness to take women seriously as scientists, or to take into account their needs to balance work and personal life, or both, has occasionally been redressed by initiatives from women themselves or the university. Whether arising from the 'bottom up' or 'top down' several typical components can be identified that substitute for formal advising or informal support structures, or both. These include: a series of regular meetings with discussion topics based on issues of graduate education; presentations of both student and visiting scientists' research, including opportunities for feedback and discussion; seminars to present research to peers; a counselor to provide advice on the requirements of graduate education as well as issues specific to the graduate experience for women; and the presence of female professors who are balancing work and personal life as role models.

The classic issues of 'systemic or piecemeal reform', changing people to fit the system or changing the system to meet the needs of a broader range of people are at issue. If the graduate education system worked for women as it does for men there would be little need for programs to make up the difference.

BOTTOM UP PROGRAMS
Understanding the origins of support schemes and how they develop provides insights into departments' treatment of women. Some programs stem from the ideas of graduate women and function initially as social movements, relying on volunteers and the commitment of a few dedicated persons. Such programs typically experience a crisis in leadership succession when its founders graduate or leave the department. It is at this point that independent funding must be raised, professional staff hired and other necessary steps taken to institutionalize the program. Programs that become dependent upon volunteer labor are always in jeopardy and unlikely to survive, given

the strong demands on female graduate students and faculty to pursue their research and attend to other academic and non-academic responsibilities.

A department's willingness to accept responsibility for maintaining a student-initiated program is an obvious indicator of its stance toward women. Several female graduate students at the University of California at Berkeley initiated a 'Re-entry' program for women taking up computer science after a gap in their academic careers. The ability of the program founders to gain seed funding from the university administration and financial support from corporations in the region for fellowships greatly improved the chances of faculty acceptance. As it gained the sponsorship of the department, the effort was broadened to include minorities and provide services to other graduate women. Female graduate students in other hard science disciplines as well as other computer science departments in the region viewed the Berkeley support system, with a professional staff member to provide counseling and organize meetings, as a model for their departments. Even so small a gesture as the program director organizing an event in honor of women receiving their Ph.D. was noted from afar.

An important type of bottom up program that we typically observed is one that has the characteristics of a social movement. Some of the most successful programs that encourage women to persist and attain Ph.D. degrees are informal in their structure and encourage affiliation among the female graduate students, faculty, and outside role models, with discussion around issues of mutual concern even when there is an invited speaker. A female staff member took it upon herself to organize a series of informal dinners: 'It was really in response to women coming to me with problems and it was obvious from their discussions with me that they were feeling very isolated. There are only 10% women in the department and once a woman gets to a research group, it's more common than not that she is in a group with no other women. At the meetings all of a sudden someone will say, "you mean it's not me!" and there's comfort in knowing that they're not responsible for either attitudes on the part of their co-workers or on the part of their supervisor.' Such informal exchange demonstrates the universality of seemingly individual experiences and keeps issues from being perceived as personal deficits. By objectifying a negative experience and its subjective consequences, the way is opened for problem solving.

Succession of leadership is a crucial factor in the continuity of a successful program, especially grass-roots programs based on graduate student leadership. A history of mutual support and achievement can help fend off a negative reaction to a program and provide a culture which is handed down to succeeding members. Program meetings walk a fine line between women feeling free to express their feelings and share their experiences, many of which involve negative treatment by male faculty and students, without degenerating into 'man bashing' and a progression of 'one upping' horror stories that become the sole focus of discussion. By keeping the focus on specific problems of graduate education and how to address them, gatherings can be prevented from deteriorating into sessions for mere negative venting of anger, without offering a positive recourse.

In the face of an often fierce competitive and confrontational stance by males that is experienced as overwhelming and deflating, women valued opportunities to present their research in a non-confrontational, supportive environment. In practice presentations, in front of women only, technical questions were not experienced as threatening. Within this safe place women developed the skills and confidence to present before a larger group outside the university. A female graduate student said, 'When I gave my presentation, the most difficult question I was asked was by a first year woman in the group. She was very sharp . . . I don't know if that question had come from a man that I would have been able to handle it as well. I was able to tell myself, "Well, you know this." Later, I then presented a paper in Washington D.C., and it went well. I got favorable comments. I just needed a little practice.' Opportunities to develop professional consciousness, build self-confidence, and counter hopelessness are especially important for those women who are the most vulnerable and at risk of dropping out.

There is a need to expand the opportunity for women to articulate their research in familiar settings, settings that are familiar in the way in which they exchange information among themselves and present ideas. This familiar setting may be different from men's but it allows them a chance to incubate ideas and gain skills and confidence that then help them in a broader range of settings, particularly those in which male-dominated behaviors are most characteristic. On this point, a female graduate student said, 'I would like to see the women have an opportunity to just get together and talk about their research. I know other women have said that the guys in their group just get together and chat about what they're doing. And [the women] have felt that they have missed out on being helped because they're not part of it.' A talk by a visiting female scientist provided an opportunity: 'Everyone gave a little tidbit of their research and I was just sweating bullets waiting for my turn. She had a lot of good questions . . . I discovered I really knew what I was talking about!' A nurturing environment is desired: 'When you're done with your paper there's someone asking you, "How did it go? How was your presentation?" You feel a sense of belonging. That someone cares about you. This is very important.' The creation of such informal venues for scientific training helps female students consolidate a professional identity within a hostile or indifferent environment.

In summary, bottom up interventions are flexible, low cost, and allow for local monitoring at the departmental level. They effect change without the help of administrators and can be customized to meet the different moods and conditions that exist in the department. Moreover, they do not burden universities with new financial costs related to special scholarships or rule enforcement. Consequently, we have found that a strategy for the success of bottom up programs explores how resources can be gained from support outside the department or university. This is a more entrepreneurial way in which to solve the problem. As the Berkeley faculty showed, it can be extremely successful in funding a re-entry program by going to donors outside the department who are interested in bringing more women into the department. Moreover, this kind of effort can help mitigate the problems of succession and raises the legitimacy of programs and their success by receiving validation and support by the 'market'.

Bottom up programs also have weaknesses. These informal interventions do not respond well to successions of key persons because it is hard to recruit someone who is willing to donate a comparable amount of energy in order to keep the ball rolling. Another difficulty is that they have few budgeted resources and thus cannot create wider-scale changes that need things like research funding, bringing in visiting faculty or setting up special programs and research tracks for women faculty. Thus, they are effective at having speedy responses but can be fragile when, inevitably, key people depart.

TOP DOWN PROGRAMS

Other programs are initiated from above: by departments, centers, university administrations and outside sources such as government research funding agencies and corporations. The programs that we identified ranged from a token yearly dinner, regular informal lunches, scheduled meetings mixing scientific presentations and discussions of women's issues, orchestrated mentoring initiatives, to an independently funded and professionally staffed organization. At the University of Washington, administrative staff were given the following mandate by their director: 'A program must be run by educators and social scientists because women professors [in the natural science and engineering] don't want to be identified with this kind of program. They want to be known for their research.' To support this initiative, a 'Student Steering Committee' composed of fifteen undergraduate and graduate students identified as their priorities isolation, competition, low self-confidence, child-bearing and child-rearing, and lack of role models. Together with the director and assistant director, both with backgrounds in the social sciences, they organized four projects to address these issues: tutoring, peer mentoring, professional mentoring, and a support group. Objectives were stated, brochures designed and printed, 'marketing strategies' developed, and events scheduled and evaluation forms created.

Some of the 'programs' were found not to be especially targeted at graduate women. For example, even though much of its content was provided by feminist consultants on gender issues in organizations, a program at an NSF-sponsored center in the physical sciences was primarily directed at men. A female participant in the discussion groups and retreats said, 'I didn't feel I could raise issues that were important to me.' The program mainly helped to broaden male participants' career goals beyond attaining faculty positions in an elite department. In other cases, even when programs did effectively target women students, they did not necessarily work as anticipated. For example, one initiative, a residence hall for undergraduate women interested in scientific careers, employed graduate women as mentors. By requiring an extensive commitment from the mentors, the program actually impeded their graduate studies. As a result, the mentors' workloads were subsequently reduced. Another program also utilized graduate women to mentor undergraduates. As a result of interviews conducted as part of this study, the program began to focus attention on the needs of graduate women, as well. Heretofore, the implicit assumption had been that a woman, having made it into a Ph.D. program, could take care of herself.

The AT&T program, an exemplar of what could be undertaken, provided stipends, a summer research position and an industrial mentor. The commitment of the program founders to demonstrating success led them to take unusual steps to insure that the 'cream of the crop' women selected for the program and its well funded fellowships, were not deterred by obstacles in their path. When an academic advisor refused to support a dissertation topic, permission was obtained to have the woman complete her research at then AT&T Bell labs. When problems with an advisor arose, AT&T mentors would pay an informal visit to discuss the issue as an ancillary part of a campus visit. More important than the money, the AT&T program placed the prestige and power of a highly respected industrial laboratory on the side of female graduate students, providing an often essential counterbalance to overcome negative experiences in their department. Unfortunately, as corporate resources for women and minority programs have declined in recent years, the former AT&T programs (now Lucent) are less well supported than in their early years.

While top down approaches offer many benefits in bringing about change in departments, they appear under-utilized. An initiative to assist female students that merely provides an outlet for women to discuss their problems or lacks administration is inadequate. Top down approaches provide a formalized means for bringing about change. Without issues being formulated and struggles made for change, a mechanism for resolving problems is lacking. A program cannot be successful without advocacy for change as was demonstrated at one university where problems were aired but not resolved. A female administrative assistant to the department chair served as a sounding board for female students, providing a place to 'vent'. The result was that students reported problems anonymously, fearing reprisals because they viewed her as powerless.

The conditions for successful program implementation were: support from above; a designated director who is not on faculty; adequate budget; help in fundraising; faculty involvement; continuous evaluation and student involvement in design and implementation. In the strongest formal program that we identified, the Director and Assistant Director played a strong advocacy role from an independent base outside the science departments. Both women made themselves available to students and had exceptional counseling and leadership skills. There was a general consensus that the qualities required of a program leader were those of a clinical social worker, rather than scientist, although a minority questioned whether a non-scientist could understand the special issues of academic science. However, on one campus, effective group meetings were initiated for women scientists by the Student Counseling Center in response to individual requests, all reflecting similar difficulties in one hard science department.

Perhaps the most important element for systematic and long-term success of a program intervention of any type is support from above. As one program founder put it, 'I started [the program] because of what I saw and because there was a positive enough atmosphere. [The chair] thought it was a good idea. Without him I wouldn't have done it. He funded it the first year.' Without such support, women may be denigrated for participating in a program. Indeed, expectation of criticism from the men in a department deter many women from participating. On the other hand, support from above creates a legitimized and safe space in which female graduate students and faculty can initiate their own projects for change. As a female scientist explained the situation: 'We were able to start the group because of the positive environment.' Speaking of an influential professor who served as the department ombudsman, a fifth-year student describing his vigorous advocacy and personal interest in women's graduate careers explained it simply: 'You must have leadership from above . . . what you really need are tenured people with guts.'

In summary, the top down intervention has the benefits of creating an incentive structure that promotes faculty to make changes by providing research funding, special tracks for women's development and special programs in funding to bring women into the department. In this way, faculty members see an economic incentive to change their behavior and one that can benefit the department more widely. The top down approach also has the benefit of legitimating the change and creating enforceable rules for persons who resist change. Top down interventions are, however, difficult to implement because of their financial and political costs. Finally, on a less tangible dimension, top down interventions force universities to admit a problem. Consequently, the added visibility of creating a top down intervention and the costs of monitoring and managing the politics often create generic barriers to their creation and success.

One way that these kinds of problems can be overcome is for administrators to go to outside agencies and organizations that rely on the university for the creation of scientists. The early AT&T program is an exemplar of this kind of activity. Not only does going to a prominent and successful scientific company like AT&T bring legitimacy to the change effort, it also tends to assuage the risk of admitting a problem because a company that relies on top-notch science has endorsed it financially and publicly. Thus, a way to make top down approaches more effective is to create networks to powerful and resource-rich organizations that can provide tangible and intangible resources to the university for supporting the programs.

IDIOSYNCRATIC PROGRAMS

Idiosyncratic interventions are neither informal bottom up interventions that rely on the teamwork of individuals nor top down approaches in which administrators bring to bear their formalized power. Rather, they occur when a single individual within the department attempts to make some localized changes or to fill a gap in the present system for the treatment of women.

As a novel strategy, a few departments seeking to achieve gender equity and upward mobility have adopted a strategy of attracting highly qualified women. In such a case it was reported that the new leadership of a department ' . . . had just gone through a revolution together, had thrown out the previous director.' There was a feeling of camaraderie and an understanding of women's issues in the department and its 'culture of inclusion' became a marketing tool to attract the best female students.

Occasionally, university administrations take direct measures, for example to encourage these kinds of strides by offering to make positions available if women or minority faculty members can be recruited. At other times, these steps have encouraged departments to hire their first female faculty member, but usually, these types of interventions rely on individuals acting on their own initiative and resources. 'We had a graduate program director who took this issue up as a personal cause,' said one interviewee who reported that it was most important to be stringent on sexual harassment so that everyone knew it was morally and legally wrong, officially and unofficially. Nearing the end of her tenure the dean regretted not having set in motion more programmatic innovations to institutionalize her personal commitment.

In another instance, a graduate student in psychology was hired for a half-time position to initiate a program and was later given a full-time staff position. The Dean purposely chose the program director from outside of the department and provided her own physical space, telephone, computer, copier, fax and graduate students as staff members, using money from the teaching assistant budget. A 'Gender Equity Task Force' was established to provide a framework for the program, with the expectation that it will be expanded to additional departments.

Often, female administrators become substitute advisors who help women graduate students negotiate political and resource constraints. One said: 'This is not a formal role that I play within the department. The students have a lot of contact with me; they've probably learned word of mouth that I can be trusted [because] I have no direct power or control. I am basically a facilitator so they don't have to worry that I'm going to cancel their RA [research assistantship].' Female administrators sometimes save women from giving up the pursuit of a doctoral degree and the chairperson can identify them as the department's 'program,' without making a commitment of resources.

Providing informal support for female Ph.D. students was an overload on these female administrators' work responsibilities, lacking in official status and recognition. An administrator said, 'It's frustrating because I don't have the power to do much about any of this . . . '. Nevertheless, a female graduate student said: 'One of the people here who really softens the blows is —. I'm not exactly sure what her position is. She'll say, "Oh, you need another week on your thesis." It's very important to have her here.' Another said, 'One thing that — [graduate student administrator] told me when I got here was don't have your academic advisor be your thesis advisor. Have at least two people.' With two faculty members to rely upon, if something goes wrong or is lacking from one, there is at least some possibility of recourse. Since the normal workings of the Ph.D. training system place individual students in a virtually 'feudal' relationship of dependency upon their professor, female departmental administrators and secretaries, in providing an underground support structure for graduate women, mitigate some of the failings of the system.

In summary, idiosyncratic changes are the most fragile because they rely completely on an individual who is acting out of a personal cause and taking an individual leadership position but without resources. Moreover, much of their knowledge and understanding of how to improve the system is not imbedded in any institutional framework but is simply information that they have on hand and that is kept only with them so that when they leave so does their knowledge. Conversely, where these kinds of programs can be successful is when the efforts of the individual become visible to other powerful individuals in the organization who then see the idiosyncratic changes as a model for changes in other programs.

STRATEGY FOR DEPARTMENTAL REFORM

A chair's leadership can influence departmental culture and climate. In some instances, however, department values and attitudes are controlled by a cadre of prominent senior male scientists whose capacity to bring money into the department overrules the status of the chair. When the values of those in power are discordant with the values of women faculty, tensions are inevitable. A 'Don't Ask, Don't Tell' dynamic can develop in which some women faculty inadvertently collude with the indifference of those in authority. After publicly being the object of overt bias by male colleagues, one lone woman chemist asserted, 'The chair is probably not aware. I don't like to bother people about things that are probably not all that important.' She went on to add, however, that in fact she is 'sure he is aware that some of the senior faculty don't say the nicest things about me or to me.' Attitudes of community and collaboration, as well as biased attitudes against female faculty members, emanate from the top down.

There is no specific point where change takes place through increase in number alone except in a very few areas, primarily in the biological sciences, where the significant number is 50% or parity. In a few traditionally female fields, women have attained majority status in some departments and achieved positions of power. When equality is reached, it usually indicates a change in power relations as well, but this can also occur at a much lower numerical level. Whether change originates bottom up or top down, intervention from above is the most salient factor to making it last. It is not necessary for reform to be initiated by the leadership of a department or the university, but it has to be supported by them to endure.

Through the improved quality of life for those women who have gravitated to relational departments, when the structure strives to support all its members, participants are freed to do optimal work. This final model, relatively free of tension, appears to demonstrate the potential for a new social organization of science. It requires energy from those who have departmental power, particularly the chair. It is vulnerable to power groups within the structure who seek to maintain the status quo. However, when a critical mass of like-minded women and male faculty feel sufficiently safe to wrestle with issues around gender, family concerns, the tenure clock, and the many obstacles which have affected the entry into science of females (and sometimes males), the scientific endeavor is only strengthened.

CONCLUSION: INTERACTIONS AMONG PROGRAM TYPES AND DEPARTMENT CHANGE

We found that informal programs, while valuable, could be extremely vulnerable to subtle and overt prejudicial attitudes without the support of department leadership. Those informal grass-roots programs which not only survived but flourished did so because of positive influence from above which then set the tone within the department. In this way a domino effect was created in which strong leadership among students and faculty members could safely emerge and creatively develop a grass-roots program in which other students and faculty members would then feel free to participate. Thus the stage appears to be set from the top down by enabling a sufficient reduction in stigma to allow for programs to develop and expand.

The issues faced by women in science in the United States are mirrored in the experience of their colleagues in other countries. Nevertheless, significant differences, both positive and negative, can be identified among scientific institutions globally in their treatment of women. This variance reinforces our conclusion that organizational and cultural factors depress or improve the number, status, and achievement of women in science.

15　International comparisons

In recent years the question 'Why are there so few women in science?' posed more than a quarter of a century ago by sociologist Alice Rossi, has been raised by her counterparts throughout the world, including the late Virginia Stolte-Heiskanen (Finland), Esther Hicks (the Netherlands), Fanny Tabak (Brazil), Mary Osborne (Germany) and Pnina Abir-Am (U.S.), among many others. Several intriguing anomalies in women's experience in science emerge from analysis of a range of contrasting national and social circumstances in the work we draw upon here. For example, female scientists and engineers in India have been found to be more productive than their male counterparts, as measured by numbers of research papers and patents produced, while Venezuelan women researchers are slightly less productive than men (Lemoine, 1994).

Women have attained greater access to higher-level positions in some southern European countries than in northern Europe (Talapessy, 1994). The nuclear family, characteristic of advanced industrialized societies, in the absence of substitute support structures, typically places a strain on women scientists. The traditional extended family, still commonplace in developing countries, provides significant support for female scientists in countries such as Brazil and Mexico.

Seeming contradictions are intertwined with unexpected findings about gender and science. Women have made the greatest gains in participation in technical fields under conditions where science is relatively low in status in comparison to other professions, for example in Turkey. A shortage of men due to their diversion into military service has also opened up scientific careers to women, in the United Kingdom (Mason, 1991) and Portugal (Ruivo, 1994). A rapidly expanding higher education system, propelled by industrialization and modernization, also works to open up scientific education, and to a lesser extent, scientific and technical careers, to women. Conversely, a declining academic economy also results in a feminization of the university as men leave for higher-paying positions in industry, especially in fields such as computer science (Lopez, 1995; Carrasco, 1995).

Using the available literature, and through interviews with researchers on women in science in a number of countries, we compare women's experience in science in developing and highly industrialized countries, in northern and southern Europe, and in socialist and capitalist contexts. We address the following questions. Is women's limited participation in science an inevitable feature of the persistence of traditional gender roles? What difference does social structure make? Under what conditions do barriers to women in science fall (or at least shrink)? The paradox of women's participation in science is that their numbers appear to increase most under contrasting conditions of system expansion and economic decline, with even the advances reflecting, to some extent, continuing inequalities among men and women in science.

WOMEN SCIENTISTS IN DEVELOPING AND SEMI-INDUSTRIALIZED COUNTRIES

In developing countries, with the notable exception of a few nations, there are many fewer women in science and engineering fields in higher education than in health, education and law (United Nations, 1995) Although cross-national data on women in universities are limited, an international research consortium of agricultural research institutes provides an interesting indicator of women's participation in scientific careers. Some women were on the non-scientific staff or at the trainee level but few could be found at the senior scientific level or in managerial posts. Nevertheless, there are intriguing anomalies such as mathematics where women can be found in university positions at higher proportions in such countries as Columbia, India and the Philippines than in many developed countries.

Turkey: Class is stronger than gender

In Turkey, the question of women in science can almost be reversed and instead of asking 'why so few?' one author has asked 'why so many?' (Oncu, 1981). The answer lies in part in the country's history of westernization in which advancement of women was part of the 'kemalist' ideology. Despite its association with modernization, science was not as closely connected to the centers of power in society as law and political science, fields in which women continued to have extremely low rates of participation. The answer is also class-based. The creation of large numbers of professional positions with the founding of the modern state meant that there were insufficient upper-class or upper middle-class men to fill them – hence the openness of technical fields to upper-class women who were typically encouraged by their fathers to pursue high-level careers.

An ideology of modernization combined with the carryover of a traditional support structure for child care, or simply the financial ability to obtain assistance, enhanced the ability of well-to-do women to pursue scientific careers. However, the attenuation of the founding ideology of the state since the 1950s, combined with the expansion of the university system during the 1980s into more traditional provincial areas of the country, has produced an unanticipated reduction in the participation of women in academic science (Acar, 1991). Nevertheless, even in the metropolitan universities where women have long had high rates of participation and access to high-level positions, there are still strong differences between men and women. For example, women report that they tend to be excluded from informal sources of communication.

Despite the ameliorating factors discussed above, women experience conflict between work and familial roles. An indirect indicator is that a higher proportion of women scientists than men are single and without children. A significant number of women, especially those who rise to high-level positions, are apparently following the 'male' model of science. Behind the façade of higher rates of participation and promotion in some sectors of the Turkish academic system, female scientists in Turkey face many of the same informal and subtle barriers found elsewhere.

Brazil: The significance of traditional gender roles
Despite differences in level of development, academic tradition, or world region, women face similar disabilities in pursuing scientific careers. A recent examination of the situation of female scientists in Brazil exemplies this conclusion (Tabak, 1993). From 1970, data were collected at five-year intervals on female participation in the hard sciences at the Federal University of Rio de Janiero. In addition, three focus-group interviews were conducted with women scientists. The number of senior female faculty members in the hard sciences at the university is negligible.

In Brazil, as elsewhere, women encounter a workplace with a rigid structure that does not take into account their need for flexibility so that they can combine career and family. Similarly, an authoritarian 'male' style of laboratory leadership, which discourages cooperation, is commonplace. In addition, women were often excluded from career opportunities, such as invitations to participate in conferences. Some conference organizers simply assume that they would not want to come since they had small children at home.

These unequal conditions in the workplace are an overlay on unequal gender roles in the larger society and a 'machismo' ideology that works against women in science by condoning sexual harassment and legitimating their lack of promotion to higher positions in the scientific community. There are also tensions created in a patriarchal society by women's occupational success. If their husbands are not as successful as they are it creates a difficulty and tends to lead to separation. In one instance a woman completing her Ph.D. dissertation did not appear at a party in her honor, as her husband had left that evening. A supportive husband sharing in the tasks and responsibilities of the household was important to Brazilian female scientists' ability to carry on research but was not always available.

Male bias toward women in science is exacerbated by women who follow the 'male' model themselves and act negatively toward female colleagues who do not follow this model. The single female member of the Brazilian Academy of Sciences does not believe that women face any special difficulties. On the positive side, the Brazilian National Research Council has taken an interest in gender issues and monitors women's enrollment, graduation rates, participation in professional careers and access to positions at universities and research institutes.

Mexico: The effects of gender socialization

In Mexico, women's participation in science increased to the level of 24.3% by 1990 propelled by the growth of female students in higher education (Blazquez, 1991). From 1969 to 1985, as higher education enrollment expanded more than fourfold, the rate of growth for women was almost three times that of men, with women constituting 44% of the undergraduate population in 1990. Although the data are not broken down by disciplines, a good indicator of the growing participation of women in graduate education is the increase from 23% to 33%, from 1971 to 1989, of scholarships awarded to women by CONACyT, the national research funding agency. Of course, these figures also show a gap between the increase in the proportion of women at the undergraduate level and the lower but still significant number at the graduate level. A gender analysis of two of the country's leading scientific institutions showed that women represented 26% of the researchers in the schools and centers of the National Polytechnic Institute and 30% of the scientists at the Autonomous National University of Mexico (UNAM). However, women represent only 2% of Mexico's scientific managers and policy makers.

In several ways the experience of women in science in Mexico is similar to Turkey. Most female scientists grew up in well-to-do, highly educated families. As children they received two cultural messages: (1) a traditional gender expectation to marry and have children complemented by (2) strong encouragement to become highly educated themselves. Parental advice to advance the knowledge of their future children placed the education message in a traditional gender context. However, by the time these future female scientists arrived at the university, the message to educate oneself had taken on a life of its own. Women expanded upon the injunction to become cultured by developing the goal of contributing to the advancement of knowledge themselves. In doing so, these women did not neglect traditional gender roles, they merely tried to make them compatible with their new career goals. This process was aided by the circumstance that many female scientists married men who were researchers themselves. Also, as in Brazil, the extended family was often available to assist with child care (Blazquez, 1996).

Women very seldom are found in high level scientific posts in Mexico. Even when women attain such positions, a man is still usually in charge and handles external relations while the woman manages the internal aspects of the organization. One reason offered for the lack of women in high positions is that they are typically not interested in engaging in the politicking required to achieve senior status. Many women, no doubt, eschewed this informal aspect of scientific advancement because of the constraints on their time imposed by family obligations. However, another factor, the particular nature of women's scientific formation, an indirect effect of discrimination, can also be inferred to play a role in their concentration on scientific research itself rather than its ancillary political and organizational aspects. To be taken seriously as a potential scientist, women had to demonstrate a greater knowledge and research ability than their male counterparts (Blazquez, 1996). To do so meant an extreme concentration on securing their knowledge base, with a concomitant effect on their style of research. Women typically develop their research findings more fully than men before publishing, a phenomenon that has also been noted in the U.S.

Portugal: The loss of males
Under certain conditions of great exigency, women's rapid entry and advancement in the scientific system has proved possible, at least to some point. In Portugal, colonial wars in the 1960s and early 1970s

removed large numbers of men from the university system temporarily, opening the way for women's participation even in disciplines heretofore largely male-dominated. By the 1980s, this enlarged pool of female graduates in the sciences had been translated into a high rate of Ph.D. production, with women the majority of new Ph.D. recipients in such fields as chemistry (77.8%), mathematics (54.5%), physics 58.8% and biology (71.4%) (Ruivo, 1987: 387). Nevertheless, women had not yet gained entry into the higher levels of the research or science policy-making systems.

One hypothesis for this difference between increase in participation at the lower levels and continued exclusion at the higher levels was women's lack of social power in Portugal. On the other hand, in industrialized countries with lower rates of Ph.D. production and presence in the mid-ranks of researchers, some women have attained high science policy positions, perhaps owing to the general increase of women's social power in these countries (Ruivo, 1987). Another hypothesis to explain the increase in women's participation in science in semi-industrialized countries is that science is still viewed as a cultural endeavor, with little relevance to the economic and political centers of power. In societies where men retain virtually total control over the levers of power, women's participation in areas of society that are considered marginal locally, if not internationally, may be unexpectedly high. This produces such anomalies as higher rates of participation of women in some scientific occupations in southern than northern Europe. Nevertheless, women in southern Europe experience some of the same disabilities female scientists encounter elsewhere, along with an additional cultural overlay of resistance to their full participation at the higher levels of science.

Greece: Traditional gender roles

The condition of women in science in Greece is influenced by two factors: (1) the traditional weak position of women in Greek society; and (2) newly emerging government initiatives for expansion of research capacities to further economic development. Although three out of five persons entering an expanded university system in the late 1980s were female, this has not yet translated into changes in the composition of research groups or participation in higher degree studies. Most women continue in traditional female fields such as languages. Although there has been some increase in women entering medicine and dentistry, these are not research-intensive fields in Greece. One hopeful sign for the future is the increased number of women pursuing degrees in mathematics, an important precondition for access to technical careers.

Greece appears to be in an earlier stage of the transition noted in Portugal. In the Portuguese case, a significant increase in female university enrollment occurred earlier in the 1960s and spread more rapidly to the sciences due to the high proportion of men called into the military. In Greece, as elsewhere, the paucity of women in high science policy positions, and the lack of programs to encourage women to pursue research careers in science and technology, retard change. Nevertheless, an underlying condition driving change is present: the need to develop human resources to make the transition to a higher-tech economy. Sooner or later, the realization that half the potential talent is not being fully utilized will help drive change and improve the condition of women in science and engineering in Greece, as it has in other industrializing countries.

There are also particular characteristics of the Greek higher education system that work against women's increased participation in academic science careers. For example, geographical mobility, a factor long noted as a prerequisite for success in traditional 'male'-oriented science systems, in the Greek context means not just relocation to another university or region as in the U.S. but typically to another country for advanced education. This higher geographical barrier exists because Greek universities have not yet organized formal graduate programs, with course work and so forth. Since national degrees are not yet taken seriously, acquisition of a foreign doctorate is virtually a prerequisite for appointment to an academic position in the sciences. This is yet another instance of a seemingly meritocratic practice working against women in science. Family pressure for young women to stay close to home is an overlay on other pressures that reduce geographical mobility. The highly politicized nature of the Greek academic and science system also works to exclude women, who are largely left out of political decision making (Cacoullos, 1991).

GENDER DYNAMICS IN HIGHLY INDUSTRIALIZED COUNTRIES: THE EUROPEAN EXPERIENCE

Despite variations in culture and politico-economic systems across Europe, female participation in the labor market and in higher education has risen considerably, yet the common contradiction of women in science and other high-status professions persists. Science follows the general rule that 'the higher one goes up the ladder of the occupational status hierarchy, the fewer the women' (Stolte-Heiskanen, 1991: 3). Moreover, despite the existence of extensive social support systems in many European countries, female scientists still face the inflexible constraints of the scientific research system, including the coincidence of child-bearing and child-raising years with the expected period of high research productivity.

There is a self-defeating dynamic at work at the intersection of gender and human resource policy in science and technology. All European countries give high priority to the production of new knowledge and the education of knowledge producers, yet they do not realize the full value from their investment. Although these human resource policies are not directly focused on gender, since women constitute at least 50% of the potential target of the policy initiatives there is inevitably an impact. The expansion of higher education during the past forty years has opened up new opportunities for women and men. Although women are increasingly being educated in formerly male-dominated fields in the sciences and engineering, improvements in access to educational qualifications have not opened up career opportunities to the same degree (Stolte-Heiskanen, 1994)

There is a lag between the attainment of equality in access to education, and its translation into jobs and especially into higher-level scientific positions. A comparative study of research groups in six European countries found that the small proportion of female unit heads made comparison of male and female leadership impossible to analyze quantitatively. The fewest women were found at the highest level and the greatest number at the lower levels of the group where 'the sex distribution is more even, or even reversed' (Stolte-Heiskanen, 1983: 65). In parallel with U.S. findings (Rossiter, 1978), improvement in women's participation in scientific research groups was greatest in faster-growing areas such as biology and chemistry and least in slower-growing fields such as physics and mathematics. One notable national difference, the tenfold greater number of women in mathematical and engineering research groups in Hungary than in Austria, both former members of the same political unit in the not too distant past, illustrates the historical variability of women in science and its amenability to policy influence (Stolte-Heiskanen, 1983: 66).

Austria

Research on women in science in Europe confirms that a 'pipeline' policy of insuring access to scientific training is a necessary but not sufficient condition to overcoming the barriers to participation of women in science. For example, in Austria women gained access to higher education a century ago but only in the decades following the Second World War, when there was a push to raise the level of Austrian science to higher international levels, did women's participation increase significantly as a result of general policies to expand the proportion of the population participating in higher education. However, in academic research settings women predominate at the lower levels as assistants but at the upper levels represent only a very small proportion (1.5%) of the directors of research units in the natural sciences (Gaudart, 1991: 18).

From the 1960s, as the result of pressure from the Austrian women's movement, the issue of women in higher education and as a topic of research and teaching came to the fore and became linked to a related debate on the role of research in national development. A new Ministry of Science and Research was established in 1970, headed by a woman, Hertha Finberg. Having a woman in a leadership policy position helped insure attention to the promotion of women at the working level of the research system. The national organization of university heads, the Rectors Conference, also directed by a woman, established a working group to remove barriers to women pursuing careers in academic science at the highest levels. Although most professorships in the sciences and most university managerial positions are still held by men, the hopeful trend is the transformation of the 'anti-feminine climate in, and masculine dominance of, academic circles' (Firnberg, 1987, quoted in Gaudart, 1991).

Finland: A 'motherhood myth'

Another small European country, Finland, also experienced an upsurge in the percentage of women enrolled in higher education in the mid-1980s, reaching 52% of the student population. Yet the proportion of women in teaching positions lagged very far behind. Females held 3% of the associate professorships in engineering in 1986 and 2% of those in the natural sciences. Women represented 4% of professors of natural science while the percentage in engineering was too low to register. Women were fairly well represented in lower-level teaching positions: 24% of the teaching assistants in the natural sciences were female, as were 15% of the engineers. The percentage of lecturers was 16% in engineering and 9% in the natural sciences. At each step of the career ladder, women are older than men as a result of time devoted to their families (Stolte-Heiskanen, 1991).

A majority of female Finnish scientists are married and more than half have children by the time they receive the Ph.D. (Luukkonen-Gronow, 1987). Not surprisingly, with each additional child, the time available for professional work decreased. Women scientists reported that, even though they had to cope with most household and child-rearing tasks, family life provided the sustenance to make up for the 'disadvantages and emotional stress experienced in their professional environment' (Luukkonen-Gronow, 1987). Beyond the general 1987

Equal Rights law, there are no specific policies to improve the condition of women in science. Although an official committee established in 1981 to address problems in women's research careers put forward a series of recommendations for increased appointment of females to research posts and provision of social services such as day-care centers for children, follow-up actions were not taken (Stolte-Heiskanen, 1991).

Some movement towards equality in household tasks has been identified but with ambiguous effect on scientists' family life. A seven-country comparative study showed that Finland had the most equal division of domestic labor. Nevertheless, research on younger Finnish women scientists found that 'women primarily bear the burden of responsibility for the reproductive activities of the family' (Luukkonen-Gronow and Stolte-Heiskanen, 1983: 273). Apparently, the women studied were well aware of this eventuality and had adopted the strategy of scheduling their first child during the writing of the MA thesis, a phase presumed to be more compatible with pregnancy, and their second child before beginning their careers. Of the scientists interviewed, 79% of the women and 51% of the men stated their belief that the main reason there were not more women in science was the difficulties that reproductive and familial responsibilties engendered for a research career. The authors suggested that these scientists were buying into a 'motherhood myth'. Nevertheless, their own data showed the obstacles that women had to overcome to maintain their scientific productivity on a par with men.

Italy: Persisting marginality

As with other European countries discussed thus far, Italy experienced a marked increase in the participation of women in higher education during the post-war era. Although women entered the initial career levels of the university system in similar proportions to men, their participation declined rapidly at the upper levels. Women also held few leadership positions in the laboratories of the national research system. A qualitative study of the work histories of sixty male and sixty female Italian researchers, undertaken in 1988, provides some provocative clues to understanding this divergence between educational and employment achievement. The study first showed that men and women scientists demonstrated similar levels of productivity in their scientific output across different types of publications (national, international, conference presentations etc.) and thus ruled out differences in research achievement as an explanation of career outcomes (Palomba, 1993).

Nevertheless, the researchers concluded that gender-related effects were indirectly responsible for the virtual exclusion of women from upper-level positions in Italian scientific institutions. Interviews with female researchers revealed that they were straining, albeit successfully, to pursue their research programs while fulfilling traditional domestic roles and meeting family obligations. Male researchers, on the other hand, were freed up by these same traditional family environments to successfully pursue their research while also having the time to engage in the 'laboratory politics' necessary for ascension into managerial positions. Like their Mexican counterparts, Italian female scientists concentrated their work efforts on their science, maintaining their research productivity at a high level, while devoting their political and managerial talents to balancing the demands of research and family. The persistence of traditional sex roles contributed to a gendered division of labor in the scientific community, largely excluding women from managerial and policy roles.

The Netherlands: A continuing dilemma
Why are there so few women in upper-level scientific and technical positions in the Netherlands? A study of the female professoriat in Dutch universities found that many of the women who have achieved high academic positions in the sciences and engineering are childless. 'Ironically . . . they are implausible role models for the potential combination of career and family. Indeed, the majority were of the opinion that a research position and a family is a difficult to impossible combination' (Hicks, 1991). The incompatibility of the 'male' model of

science, with its long hours, and the policy goal of opening up scientific research careers to larger numbers of women comes up against the strictures of the traditional sexual division of labor. Combining two careers with stringent demands – scientific research and motherhood – is a difficult task. Only a few women may be able or willing to pursue both roles simultaneously, at least as they are structured at present.

Men's long hours in the laboratory are made possible by female responsibility for the 'private sphere'. Women's research time is truncated unless they give up that private sphere. Since most women are unwilling to do this, the prospect of overcoming their small numbers at the top is highly dependent upon ' . . . erosion of the norm that women have sole responsibility for family and household maintenance' (Hicks, 1991). The restructuring of the scientific work role, making the emerging 'female' model (limiting time at the workplace) the norm would also appear to be a prerequisite for change. Other partial alternatives include Alice Rossi's idea for a 'technical fix', professional household care firms that remove some of the burdens of home maintenance from both men and women. Expanded child-care facilities would also reduce the ' . . . alienating choice between home and profession' (Hicks, 1991: 186). Perhaps ironically, it may be easier for an upper middle-class female scientist to pursue a demanding scientific career in a Third World country where a personal support structure of extended family and servants is assumed. Indeed, the most difficult career point for many female scientists from developing countries is the years they spend in a developed country, bereft of such assistance.

Israel: A few women at the pinnacle

Despite obstacles in their path, a small number of women do attain the highest level of formal position in academia: the full professorship. In Israel, where traditional expectations of female responsibility for child care are strong, a recent study found no diminution of scientific productivity, according to such measures as number of papers published, due to combining a demanding career with family roles

(Toren, 1991). Approximately two-thirds of the Israeli female full professors were natural scientists. Among the explanations for the lack of negative impact of heavy child-raising responsibilities is, of course, the fact that the smaller number of female full professors may have higher abilities than their larger number of male counterparts. Child-raising was noted to decrease participation in international meetings and the ability to take advantage of fellowship and research opportunities abroad, both activities especially important to advancing a scientific career in a small peripheral country. The unexpected high productivity of these female full professors demonstrates the invalidity of myths that propound an inherent contradiction between demanding scientific work and marriage and family.

Denmark: The difficulties of balancing

Although many women scientists develop creative strategies to overcome the handicap of burdens placed upon them by their having primary responsibility for raising children, this should not be taken to mean that they do not experience career and personal difficulties along the way. A recent study of tenured female scientists in Denmark illustrates the impact of motherhood on a successful scientific career. A first-order effect, similar to the one identified by Cole and Zuckerman among female scientists in the U.S. (1987), is that ' ... they sleep less and skip many leisure activities' (Nielsen and Elkjaer, 1991). A second-order effect is the loss of research opportunities because their work schedules are 'reduced to normal working hours'. In contrast to the Israeli cultural pattern of having children early, some female scientists in Denmark report that they delayed having their first child and ascribed part of their career success to their ability to arrange child care. In Denmark as in Israel the traditional ideology of the family is strong, with women having primary responsibility for maintenance of the emotional relations of marriage as well as more mundane household tasks. These Danish women scientists, who aspired to successful careers and family life in tandem, had to juggle the demands of both research and motherhood in contrast to their male counterparts whom, they felt '. . . can better devote themselves to research, because they are not mainly responsible for the family'.

WOMEN IN SCIENCE IN SOCIALIST COUNTRIES

What difference does a socialist system make to the position of women in science? As in Austria and Finland, the former German Democratic Republic experienced a similar sharp increase in females educated in science and engineering after the Second World War. However, the increased flow of women through the 'pipeline' did not readily translate over time into anywhere near similar proportions of women occupying leading positions in scientific institutions. Even when their numbers grew close to parity (40%), this was not reflected in attainment of significant numbers of higher-level positions. Greater numbers, by themselves, did not bring about change. Nor did the availability of social support such as child-care facilities and generous family leave policies. Instead, the persisting patriarchal culture of the scientific workplace, with inappropriate patterns of communication and work organization for women, was identified as the problem that impeded career advancement (Radtke, 1991).

Similarly, in Bulgaria, where access to a scientific career improved greatly during the post-war era, coincident with the period since the Socialist Revolution, nevertheless there remained the overall delay in the promotion of women within the scientific system. Also, a gender division of intellectual life persisted, including a continuance of a traditional association of women with teaching. Female scientists report that they face a heavy load of family duties, with some assistance from their mothers for child care. Again this household burden makes it difficult for women to combine scientific achievement with administrative advance. Two out of three roles appear to be possible in tandem: for female scientists it is typically research and family; for males, research and politicking. Even when women achieved some measure of advance into higher-level policy positions within the Bulgarian Academy of Sciences it was typically accomplished through election to the position of scientific secretary, a role in accordance with the traditional domestic supporting role of women (Ananieva, 1991).

In the former Soviet Union the proportion of female scientists reached 40% during the 1970s and 1980s (Koval, 1991). Indeed women represent 58% of the engineers and 67% of medical doctors. Nevertheless, ' . . . there is a hierarchical difference in the division of labor between the sexes.' As found elsewhere, men monopolize the decision-making positions while women predominate among the second level as assistants. Again family duties and child-rearing are an overlay on career responsibilities. Women are also viewed by men as less able to do science with the consequence that their lack of career advancement becomes a self-fulfilling prophecy.

Participation of women in scientific research groups was found to be highest in socialist Hungary and Poland where the integration of women into research positions, if not promotion to leadership roles, was a general policy tenet. Women attained high rates of participation in many science and engineering fields in Eastern Bloc countries during the socialist era but are losing their positions at higher rates than men during the post-socialist era. However, even during the socialist period when official ideology prohibited direct discrimination, female scientists typically filled the middle ranks of support researchers working under the direction of a male laboratory chief.

Despite significant variation in the organization of scientific institutions, differing socio-economic systems appeared to have little effect on the condition of women in science. In both capitalist and socialist countries, women were deterred from promotion to the managerial and policy levels of science and found that their family life provided satisfaction that at least partially made up for problems in work life. There was a common deficit of female scientists in the higher levels of research organizations. The exception to this rule in both systems was associated with the position of science in society. Even when, as in the former Yugoslavia, women achieved significant positions in scientific research institutions, this advance was associated with a decline in the position of science in society. 'As it became less and less prestigious, science opened up to women' (Blagojevic, 1991:75).

MOVEMENT TOWARD CHANGE

The situation of women in science in Spain exemplifies the principle that women make significant advances in a rapidly expanding system. In the later 1980s, a period of substantial growth in R&D investment, the number of women researchers increased by 180% in comparison to an increase of 88% for male researchers. Female researchers were, of course, starting from a much smaller base but, when opportunites expanded, it is interesting to note that women actively took advantage of them to pursue careers in research science. As in other countries, the proportion of women in research training programs is significantly higher than the proportion in research positions. Thus, it will be possible to expand upon the successes of the late 1980s when women's share of research positions grew fom 22% to 26%. Whether such expansion will actually take place, without a continued increase in research funds is debatable.

The entry of women into the engineering professions in Spain has not taken place without some ironic contradictions. A tightening of admissions procedures led to an increase in the proportion of women in telecommunications engineering. Although their numbers were still low, some improvement was generated by policies that limited places and favored secondary school students with higher grades. Since the relatively small number of women tended to get better grades than men in science subjects, '...whatever the original intentions behind the admission policy, the priority given to overall academic criteria has in fact led to a feminisation of the student body enrolled in university technical programmes' (Alemany, 1991: 219). Despite academic harassment by men, women do significantly better than men in examinations and show a 'greater commitment to [their] studies' (*Ibid.* 224) in electronic engineering, in part, they are impelled to achieve in order to counter the effects of discrimination.

The rapid increase in educational attainment of women in science in southern Europe has given rise to expectations that there will be a quick throughput into the staffing of the academic system. However, the experience in northern Europe where numbers of trained women have been available for a longer period of time suggests that this will not necessarily be the case, at least not without some intervention to open the system up to fuller participation by women. The gap between men and women in science in both Germany and England expands at the point of entry into the first real academic position after post-doctoral training (Osborn, 1993; Moxham and Rogers, 1993).

THE U.K. EXPERIENCE

In the United Kingdom, this bifurcation point leads to the formation of a dual track system of independent and dependent tracks in academia. A higher proportion of men enter the independent track where they attract research funds, grow research groups, and pursue their own research interests. A dependent track, existing in parallel and symbiotic association, has a higher proportion of women. Members of this lower-status track are limited to short-term posts where they become dependent on the research funds of others as subordinate members of their research teams (Moxham and Rogers, 1993).

Women are marginalized by the persistence of these ' . . . gender-related hierarchical structures' that inhibit them from pursuing independent scientific careers. Reistance to women increases with the height of the academic ladder. More than twice as many men as women become senior lecturers and, as late as 1991, there were no female full professors of chemistry (Mason, 1991). The intensity of feeling against admitting women into the inner circles of science can be seen in the decades-long struggle to open the Chemical Society to women (Mason, 1991). This battle was finally won in 1920, after women unmistakably demonstrated their competence during the First World War by filling positions formerly held by men.

Nevertheless, covert resistance to women in science persists to this day and is expressed in the drastically lower levels of women in high academic and science policy positions. For example, only 22 of the almost 500 professors of biology in the U.K. are female. The issue of paucity of women in senior positions, even in fields such as biology in which women had achieved significant representation, was taken up in *The Rising Tide* (Lane *et al.*, 1994), a report identifying actions to allow women to realize their potential in science and technology in Britain. Following upon an earlier science policy report, *Realising Our Potential*, recommending the increased utilization of science and technology to enhance industrial competitiveness, *The Rising Tide* also linked its recommendations to an economic theme.

Not only did the loss of female technical talent impede 'national wealth creation' but continuing gender bias was also counter to ideals of equity. The authors proposed a dual strategy of a push from below and a pull from above, modeled upon successful initiatives elsewhere, as the means to accomplish change. Thus, the Committee suggested extending to England the Scottish format of secondary education that, on the one hand, did not require early specialization, and on the other, placed science courses in a broader context. Both policies had been found effective in encouraging a larger number of young women to take the preparatory educational steps toward scientific and technical careers. *The Rising Tide* also proposed that government set targets of 25%, introducing external pressure on academic and governmental institutions to promote women and include them in policy-making bodies in signifcant numbers.

Many policies that would improve the condition of women in science are also applicable to other professions and, indeed, all working women and men. Expanding U.K. child-care facilities and teleworking opportunities were among the proposals with such broader implications. Proposals of even a prestigious government sponsored commission are likely to remain just that, unimplemented recommendations on paper, without a build-up of political pressure and highly publicized protests from those most affected and their supporters. One recent example is the 'Oxford Revolt' in which business as usual in the distribution of academic positions was rejected by female academics who protested in favor of specific measures to increase women's participation in the professoriat of Oxford University.

Follow-up to the *Rising Tide* report suggests that the issues of women in science will not go away without steps taken for their satisfactory resolution. A 1995 Forum on Gender Policy for British Science, sponsored by the Science Policy Support Group in London, signaled that the issue of women in science has been raised from the isolated concern of a few pioneers to a general matter of science policy. Nevertheless, government attention has its limits. The representative of the Conservative government at the conference expressed interest only in policy measures that could be undertaken at no financial cost. However, the broader significance of the report and associated events is that women scientists are organizing themselves to represent their interests. One of the leaders of the Committee on Women in Science, Engineering and Technology (SET) and chair of its working group, Cambridge Zoologist Dr Nancy Lane, prefaced a call for new policy ideas with a report on her meeting with the then British Prime Minister, John Major.

The ability of female scientists to access the highest level of political leadership is an indicator that the issues of women in science are moving to center stage in U.K. science policy making. The Forum participants came up with a range of ideas to build upon the recommendations of *The Rising Tide*. Valerie Ellis called for the systematic introduction of networking and mentoring in the workplace to assist women scientists to overcome the constraints that tend to limit them to lower positions. A call was made to take gender differences into account in the teaching of science in the schools by giving science more social context; girls want to understand the social contribution while boys tend to be satisfied with abstractions. The importance of organizational leadership in improving the lot of women in scientific institutions was noted by Jan Harding who presented an example of success due to a department head's commitment to change.

The fate of these and other ideas to improve conditions for women in science in the U.K. is still in doubt. Nevertheless, the issues have received a recent spate of legitimation from policy makers such as William Waldegrave, the cabinet minister responsible for science and technology. Based upon such statements as Waldegrave's that 'It is obvious that we are not using the resources of half of our people properly', they are now part of the accepted repertoire of science policy issues (Dickson, 1993). Gender and science policy has also become the topic of academic seminars and public policy meetings well beyond traditional feminist circles. Indeed, there is now a recognized intersection between the two discourses. This is a significant advance over the polite disinterest expressed just a few years ago by science policy groups who, until recently, exemplified Nancy Lanes' statement that 'One problem is that men may not entirely understand the barriers that women scientists face.' It can no longer be said that female scientists have not publicly brought the issues of gender and science to the attention of the U.K. scientific community and its sponsors in industry and government.

A series of studies and reports produced by international, multinational and non-governmental organizations, such as the United Nations, the European Union and the Latin American and Caribbean NGO Forum, has also heightened the international visibility of the inequalities among men and women in science. In addition, individual scholars, both natural and social scientists, in various countries through their scholarship and advocacy have provided the data and analytical frameworks for these broader efforts. Organizations of women scientists and technologists such as the Third World Organization of Women In Science, The Association of Women in Science in the U.S. and Women In Science and Engineering (WISE) in the U.K. strengthen the social ties of their members and represent their interests. It is increasingly recognized that the issues of inequality of women in science and technology transcend gender and equity interests, since all the human resources in a society must be used to the full to achieve economic and social development.

16 Athena Unbound: Policy for women in science

'For women there are undoubtedly great difficulties in the path, but so much the more to overcome,' exulted Maria Mitchell (1818–1889), the first female professor of astronomy in the U.S., at the then newly founded Vassar College. More than a century later, on the 175th anniversary of her birth, women's scientific aspirations are still restricted by 'tradition and authority' (Enna, 1993). Few female scientists are as happy as the nineteenth century astronomer Mitchell to have put up a brave front in the face of a host of gender-related problems. Despite significant improvement, especially in the numbers of women who wish to enter scientific careers, many of the organizational structures of science, and some scientists, continue to resist women's full participation. Additional steps are necessary, beyond encouraging women to take up science, to insure that science is open to them (Lovitts, 1996).

In this final chapter, we discuss the policy implications of our findings in terms of motivators and inhibitors that shape the experiences of women and conditions of inequality in academic science. First, the arguments behind each thesis are described and its accomplishments and shortfalls reviewed. Second, we argue for new policy recommendations related to department reform, the factor that is most relevant to the organizational issues explored in the preceding chapters and that we have identified as a means by which programs for change can be implemented.

Five strategies have been suggested during the past thirty years to promote improvement of the condition of women in academic science. These include: calls for equity, expectation of personnel shortage, national economic competitiveness, generational transition and departmental reform. Equity, of course, is based on the moral grounds that women were excluded in the past and as a matter of right they should be included in the future. Shortage, or expectation of too few recruits to science and engineering, engenders initiatives to broaden the recruitment of scientists. This typically leads to consideration of drawing more women and minorities into the scientific enterprise. Economic competitiveness or the realization that science is increasingly closely connected to developing future high technologies is the mirror image of the 'shortage' theme.

Failure to make full use of the talent in the national population creates a potential economic deficit, especially in comparison to other countries that make fuller use of their available talent pool. Generational change is based upon the expectation that a new generation of male scientists, feeling some of the same pressures to take responsibility for home and family life, will support change to realign science with changes in family structure. Departmental reform is the injunction to 'solve one's own problems' before being forced to do so by external authorities. Leadership by the department's power structure is essential to this approach.

A MORAL AND LEGAL IMPERATIVE

Laws prohibiting discrimination are on the books of most industrialized nations, and although they apply to academia, they are seldom vigorously enforced (Carson and Chubin, 1992). From the 1970s, efforts to increase the number of women in U.S. academic science departments have largely resided in affirmative action programs, requiring full consideration of female and minority candidates. However, in the 1980s lack of vigorous enforcement reduced the spirit of the law into a bureaucratic requirement that became a routine part of the paperwork of the academic hiring process, often with little or no effect on recruitment and no impact upon retention.

Nor did this strategy, focused on getting entrants into the system, address the hidden inequities of academic departments. Many academic scientists and engineers are aware that the Defense Department's strong interest in military contractors hiring women made a significant difference. Indeed, a dean of engineering proudly pointed this out as an instance of women's advancement in engineering. But when asked if the same measure could be applied to the academic environment with similar results, he admitted the 'logic' of the suggestion but expressed a lack of interest in the experiment.

Affirmative action creates a formal procedure that helps insure that a broader range of candidates are considered for a position but often without the expected result of a significant number of positions filled by women and minorities. Despite the presumption that minorities and women are insured positions through this process, a selection committee always has the option of certifying that its choice was best qualified. Since women and minorities often follow careers that are lightly off the beaten track, it is easy to find fault with their career choices. For example, a senior male respondent questioned the hiring of a female physicist by a prestigious research institute on the grounds that she had deviated from the normal academic career path by becoming an astronaut.

Most diversions from the track are much less spectacular but even accepting a less than optimal post-doctoral fellowship because it is close to home can be the basis for exclusion from a short list. Thus, the effects of affirmative action procedures in promoting the interests of minority and female candidates are debatable. A respondent said, 'I have never seen affirmative action in the sciences, never seen it achieve equal employment.' He questioned whether implementation of affirmative action policies at his university was much more than a bureaucratic façade in which invitations were issued for job interviews to minority and women candidates with little real intention of recruiting them.

Affirmative action is a double-edged sword that creates a double bind for women and minorities in the sciences and engineering. It modestly increases the numbers in the system but then is used to denigrate blacks and women and keep them down. As one respondent put it, the implicit and sometimes explicit message is that 'if we weren't being good guys, you wouldn't be here.' Those few minorities and women who do get into the system are presumed to be of lesser worth than their peers. The most subtle and insidious aspect of the exclusionary process for faculty women at all levels is the virtually inevitable stigma. When women attempt to initiate mechanisms to achieve symmetry with their male colleagues, it is used against them. The most glaring example is attributing a faculty appointment as a 'quota' hire.

An unintended consequence of efforts to empower women is that they may provide ammunition for devaluing women as needing special support. Thus, providing a mechanism for affiliation, a program can instead encourage attrition. Support of programs by senior faculty members deters this: a program that is legitimated by an authority figure escapes the presumption of deviant behavior that is often the fate of women in graduate departments who suffer reprisals for engaging in even the most modest collective activities. An initiative that is protected by an individual or group with the power to advocate or at least accept change can also be the first step to creating a supportive ethos for students in a department. Such experiences have led to a re-valuation of affirmative action among many of those it was designed to assist.

Affirmative action has at times been transformed from a public intervention on behalf of women and minorities into a private instrument used against them by some of the very people whose discriminatory practices it was designed to thwart. This experience has, not surprisingly, driven some minorities and women to the conclusion that the existence of affirmative action procedures does them more harm than good. They themselves become opponents of such policies, and the unlikely allies of the very persons who have turned affirmative action into a discriminatory weapon against them.

It is important to note that affirmative action is currently in a retrenchment trend that may permanantly wither the system. A good example is the present ruling of the regents of the University of

California to abolish affirmative action in recruitment and hiring. The system's visibility and prestige will surely stimulate mimicking by other universities which now have cause to view the system as delegitimated or politicians who will seize the opportunity to eliminate affirmative action for their political gains rather than the gain of the wider system. Furthermore, affirmative action has a systematic problem: it does nothing to affect the 'feeder pool' of qualified applicants. Because it is designed to equalize opportunity for individuals with the right credentials, it suffers from an inability to insure that persons from minority groups will get credentials in the first place. This is a central problem because it gives rise to the argument that affirmative action promotes the hiring of unqualified minorities, which in turn, defines them as 'quota hires' and intensifies feelings of animosity by the majorities.

At best, affirmative action is a necessary but not sufficient condition for inclusion of minorities and women in the sciences. It has too often become a method of protecting an academic department from protests against its lack of diversity. It allows the appearance of a neutral double-entry bookkeeping system in recruitment efforts: an overt procedure with white males and women and minorities interviewed, covering a hidden decision-making process in which white males are typically offered positions. A department head at a major university reported that his wife, who was head of the County Equal Opportunity Employment Commission, did not like to investigate academia because its employment practices were arcane and difficult to fathom. Certainly, a reinvigorated affirmative action effort and enforcement of existing federal laws could make an important difference to the advancement of women in science (Chubin and Robinson, 1992).

THE EFFECTS OF SHORTAGE

Recruiting from underrepresented groups, such as women and minorities, is always suggested as a potential solution whenever a shortage of scientific and engineering personnel appears. The First and Second World Wars opened up opportunities for women that largely closed down with the conclusion of hostilities. A human resources deficit became a national policy issue in the U.S. during the Cold War when availability of sufficient technical personnel was viewed as essential to national security (Pearson and Fechter, 1992). The issue arose most recently in the US during the 1980s.

In the mid-1980s, Erich Bloch, then director of the National Science Foundation, called upon social science knowledge to help resolve the dilemma of too few women in science. The then director of NSF's Sociology Program, Dr Phyllis Moen, was asked to analyze existing research on women and science for clues (Moen, 1986). In addition, NSF commissioned new studies (including some on which this book is based) to fill gaps in knowledge through its research funding programs. NSF also established a number of programs to improve scientific education for girls and to encourage female scientists by making fellowships and visiting professorships available.

In the short term, an open immigration policy was put in place as a temporarary solution to expected technical personnel shortages. This policy was retained during the aftermath of the Cold War in part through bureaucratic inertia but also at the insistence of 'high-tech' employers seeking open access to international labor markets. Combined with a surplus of scientists and engineers in some fields produced by the decline in military and military-related research and development spending during the post-Cold War era, the policy impetus to address the issue of unequal gender participation in science has largely dissipated. Projected shortages of scientists and engineers turned into an oversupply, exacerbated by some overly pessimistic estimates that, coupled with highly optimistic retirement expectations, led some university presidents to predict a Ph.D. shortfall that has yet to occur.

One shortcoming of policies for immigration and recruitment is that they are applied on a temporary basis: they are put into place only after a shortage problem has been identified. This lag disconnects the benefits of the policies (ample supply) from the periods in which they are needed most. Also, this policy ignores the dictum that 'supply creates its own demand.' As such, in countries such as India, Israel, and China, the investment in scientific training programs has stimulated growth in scientific as well as technical and entrepreneurial fields that spring up when the demand within science wanes.

Deficits in women's participation in science attract heightened attention from policy makers during periods of expected personnel shortages, such as the mid-1980s (Moen, 1988). A cyclical pattern can be identified, with alternating themes of equity and 'national needs' utilized to justify efforts to improve the condition of women in science (Office of Public Service and Science, 1994). When the moral injunction of equity is enforced by self-interest, the way is open to change, especially for groups that are currently underrepresented (Campbell, Denes and Morrison, 2000).

ECONOMIC COMPETITIVENESS

The third thesis to advance the cause of women in science is one of economic competitiveness. Highly trained human resources are the key to future economic development. A society that neglects the native talent in any sector of its population does so to its detriment. In the emerging information economy, universities and other knowledge-producing and disseminating institutions are the ultimate source of wealth, just as mineral deposits and ports were in the past.

So much of the prospect for future economic advance now rests on knowledge that any nation that does not utilize all its talent runs the risk of falling behind other nations that do.

Until the present the U.S. has been the greatest repository of talent in biotechnology and software, providing the basis for leadership in those industries. As other countries gear up their universities to produce more Ph.D.s locally, the need to study abroad will become less pressing. Science, of course, will remain international, but in the future, U.S. Ph.D. graduates will be just as likely to be found in attractive post-doctoral positions in countries such as Sweden as will their Swedish counterparts in the U.S. Indeed, the current abundance of research funds in Sweden, which is upgrading its ability to translate research into economic development, is proving a strong attractor to junior researchers elsewhere to relocate. Swedish professors are also importing intellectual guest workers, including underemployed American post-doctoral fellows, in order to carry out their expanded research programs.

Certainly, it is easier and perhaps cheaper for technology entrepreneurs to recruit from abroad, rather than address the long-term causes of this shortfall in technical human resources at home. Beyond the sometimes narrow perspective of individual firms and universities eager to attract the best current talent from whatever source it arises, there is a longer-term dynamic of scientific mobility to consider. Such flows of people and talent can reverse directions. This fact should give pause and lead even those companies that wish to encourage immigration to take a longer-term view. Even in the medium term, many scientists and engineers who stayed in the U.S. after completing their education to teach in U.S. universities or work in high-tech companies are returning home, often encouraged by special schemes to attract them back to head up new research units in Singapore or found companies in Taiwan.

As the international mobility of technical labor becomes reciprocal and more equal, any country that excludes a portion of its population from scientific or technical careers, whether de jure or de facto, will operate at a disadvantage. In Japan, for example, it is expected that the approaching demographic decline will produce an opening for women to enter scientific and technical careers, from which they have heretofore been largely excluded. This population problem, coupled with a national science and technology strategy that increasingly relies on an academic science base to enhance high-tech industrial development, creates an opportunity. Nevertheless, demography only creates a possibility. Without the necessary social and political impetus to translate this into schemes to recruit those formerly excluded, and modify scientific institutions to make them more attractive and welcoming, the status quo will persist.

Each country must focus on increasing the participation of its population in science and engineering in order to maintain national competitiveness in high-tech industries. Hoffman and Novak (1998) recently argued that '[If] a significant segment of our society is denied equal access to the Internet, U.S. firms will lack the technological skills needed to remain competitive.' If such a conclusion is warranted for Internet use, then surely it should be applied to the users, with appropriate implications drawn for the training and employment of minorities and women as scientists and engineers.

GENERATIONAL CHANGE

The thesis of 'generational change' is based on the assumption that a younger generation of male scientists, facing some of the same family pressures as women, will help reform the academic structure in ways that benefit both women and men. Institutional reform that takes into account the needs of female scientists also benefits men by introducing some flexibility into career paths. The traditional structure was that an individual scientist devoted themselves to their career completely, without much attention to the details of life. A spouse normally would be at home and take care of the private sphere. This allowed an individual, virtually always male, to spend 70, 80 or 90 hours per week in the lab. They could establish their lab, build their career and have a firm basis for their later stability that fit with the model of the family and the model of work structure that was then universal in society.

What may be happening at present is that the traditional model is eroding in the U.S. The women's movement has encouraged more openness to dual career families, with men having to share some of the burden of housework. At least there is a cultural movement in this direction, even if it is not always in place in fact. The previous model virtually precluded women who wished to have a family from assuming professorships, given the demands of the position. It was very difficult to combine home responsibilities and an academic career. Recently, an awareness has grown that the traditional model not only discriminates against women but gives rise to complaints from younger male academics as well. They are feeling much more pressure and, in part, it is because their own home situations are no longer organized according to the traditional model.

This introduces a stress into the system that did not previously exist. This stress is related to deeper changes in family structure and in the structure of work that has brought women into the work force in unprecedented numbers during the past three decades. Thus, there may be a shift from a single focus on strictly career issues, not just for women who are trying to move into the system, but also for younger male academics who find it intolerable to try to manage both the traditional academic jobs and their new home responsibilities. This may create pressures to modify the system in a way that universalizes it a little more than in the past. There is some evidence that this shift is beginning to happen in the U.S. To the extent that it does take place it is not only a gender issue but also a generational cultural issue.

Nevertheless, the thesis of generational change has its limits. It certainly exists on the level of wishes expressed by younger scientists, both female and male, for a less pressured way of life. However, the pressures for publications, for raising funds, are becoming even greater as competition increases, especially at the highest-status universities. This is not due to a decline in research funding, which has indeed recently been increasing, especially in the biomedical sciences. But there is increased competition for these funds, from additional universities in other parts of the country, trying to establish themselves as research centers. As scientific research is seen as increasingly closely related to future economic growth, competitive pressures increase, even though some of the most competitive people, especially among the younger generation, would like to see these pressures reduced. Despite their wishes, as the competition increases, they are forced to write more grant proposals and feel pressured to spend more time at work. These young scientists feel themselves under an even greater strain even though they would prefer a different system (Etzkowitz, in press).

The stresses revealed in the coevolution of the institutions of family and work may also recreate new and possibly greater pressures towards inequality if not managed properly. University scientists are currently experiencing intensified budget constraints that put even more pressure on research productivity and premiums on time. Universities have also placed new demands on faculty time in regard to teaching quality, widened access to students for faculty time, new electronic teaching materials, web pages, Internet class discussion groups, and university-wide published teaching assessments. All of these trends intensify the risks of starting and successfully managing an academic career, which may be more likely to 'double bind' Athena, rather than unbind her.

Another factor related to the coevolution of family and work is that professions that have increased the number of women often suffer a decrease in status and mean pay because of an increase in the effective labor supply. If this happens without an attendant increase in the demand for services, it is likely to decrease the incentive to pursue a scientific career in the first place. However, there is a counterbalancing force that may improve the condition of women in science, the opening up of new opportunities and positions in firms based upon academic science.

University science is becoming more entrepreneurial as evidenced by the number of academic scientists who are starting their own businesses, taking equity stakes in part-time businesses, or partnering with their university to start businesses that share university resources (Powell *et al.*, 1996; Murmann, 1998). Consequently, in creating firms that span the boundary of the university into the private sector, it is likely that models and accepted practices in the private sector will begin to be imported into academia. This suggests that the better gender balance (demographically speaking) and the stronger expectations for equal treatment and pay for women in the private sector are likely to flow over into academia. Thus, while these coevolutionary pressures lack the intentionality and organized features of the movements are most likely to accelerate positive change, the growth in the private sector and the power of the commercial market to pick winners will offer women more opportunities to use their skills.

ORGANIZATIONAL REFORM

A thesis of this book has been that macro-economic and social indicators tell only part of the story of disadvantage for women. Through triangulation of theory on professions and inequality, original fieldwork, and statistical analysis we have found that much of the process by which disadvantage is created and reinforced occurs within organizations and at the level of the department. It is at this level that recruitment, socialization, learning through networks, and access and referral benefits are created and combined with human capital factors. In this sense, we follow similar sociological literatures to the conclusion that the seat of change is at the department and organizational level, rather than in the environment or in human capital, because it is the organization that connects human capabilities with environmental requirements (Baron and Bielby, 1980; Baron, 1984).

RESISTANCE TO CHANGE

What are the forces for change to achieve gender equity in science and the countervailing forces for stasis? The organization and culture of academic science deters many women of high scientific ability from making their contribution. In those instances where a department has faced up to this situation and altered its behavior, women's participation has improved dramatically. A broader recognition of the need to change and requisite actions are required to reconstruct male-gendered science and engineering departments. Indeed, the experience of In-balance Program at the Center for Particle Astrophysics, University of California, Berkeley suggests that many of these changes are necessary for both women and men. Male participants in the program's seminars and retreats expressed interest in reducing the all-consuming pressures of scientific work-life, although they could not see how this could take place within the strictures of the existing system.

MARGINAL DISADVANTAGES

The succession of impediments to the entry of women into scientific careers that narrrows the stream to an extremely small flow at the stage of graduate training has been conceptualized as cumulative disadvantage. However, even given these disadvantages significant number of women receive degrees in science at the BA and even the Ph.D. levels. Nevertheless, fewer pursue careers in science and there are few senior women professors (Moen, 1988). The disadvantages that accumulate to narrow the flow into the science career pipeline are supplemented by additional disadvantages, at the margin, that discourage even the most highly motivated women who have taken steps to pursue scientific and engineering careers at the doctoral level.

Removal of some or all of these barriers at the doctoral, junior and senior faculty levels could have an effect, in the short term, in increasing women's participation in science and engineering. Taking such steps could also provide role models to assist in long-term efforts to lower barriers at the early stages of the life course cycle, thereby increasing the flow into the science career pipeline. Thus, the importance of focusing policy intervention at the later stages is twofold:

1. Encouraging the creation of a critical mass of women faculty members in academic science and engineering departments that, in and of itself, has an effect in changing academic cultures and, by implication, lowering barriers for future generations;

2. Revising the image of high-level careers in science and engineering for women from anomalous to 'normal' thus providing the incentive of examples of achievement to encourage younger women to break through the barriers prevalent at early stages of the career. As we have sugggested above, sympathetic male faculty members can play an important role in mentoring women, relieving some of the pressure on overburdened women.

These graduate students and professors, after successfully negotiating the numerous barriers to entry that exclude so many other women, often pursue less demanding careers than their male peers. These women are not lost to science. Rather they are women who, with a few exceptions, are excluded from positions in the top academic departments in their field. Many pursue research careers in industry; others have taken appointments in teaching colleges. Whether these scientists are excluded from high-level academic careers through discrimination by academic departments unwilling to accept women as equals, or for other reasons, the result is the same. There is a pool of women scientists working in industry and lower down the academic ladder whom their advisors, usually men, agree are the equal of their male peers who are pursuing research careers at the highest academic levels. If professorial jobs were made available, qualified women scientists could be recruited to create a critical mass of at least three women in each leading academic department. This would provide the range of female role models necessary to bring forth an enlarged next generation of women scientists.

Women should be recruited into 'pivotal jobs' or 'linking positions' in order to formally increase their social capital. A pivotal job is one that places someone in a position with a lot of crosstalk between other faculty members, particularly crosstalk that is often not spoken about openly but is critical to promotion and understanding of how the system rewards performance. For example, in academic circles, a position on the personnel and review committee offers great insight into how the university operates and reviews performance. It also permits members to gain first-hand knowledge about how outside letters (letters from external referees) are written, from whom outside letters should be solicited, and how issues of research productivity, teaching, and service are balanced in tenure decisions.

If women were consistently placed on such committees it would (1) widen their network of personal contacts, (2) allow them to display their competencies, (3) increase their access to information on how promotion processes work (information that is normally only circulated among male ingroups), (4) help them demystify the tenure process in their own minds, and (5) position them as knowledgeable colleagues among their peers who will in turn rely on their counsel. Some universities and professional schools have adopted this system with great success by creating 'untenured observer' positions on their tenure and review committees. These positions are open to junior (untenured faculty) for 1-year stints. The untenured observers have access to the entire decision-making process. They attend all the meetings, read all the outside letters, observe decision-making processes (and tenure battles), and learn how the decisions emerge.

At one university familiar to the authors, the expressed public mission of the observer is to reveal the scepticism felt by outsiders about the process, and report back to peers and senior colleagues on the way the system works. Another method of expanding women's access to social capital is through arranged mentoring. While it is commonly understood that mentors offer advice and access, it is less commonly recognized that assigned or organized mentoring programs can have effects that are comparable to informal mentoring. In a number of universities for example, new faculty members are assigned to teaching and research mentors. These individuals are usually just a year or two ahead of the new recruits. Their mission is to familiarize the new faculty members with the idiosyncrasies of the department and university. They pass down factual knowledge, advice, tips, and strategies that help bring the new faculty member up the learning curve more quickly.

Importantly, much of this information is in the form of tacit knowledge that is rarely available through other means but critical to success. It could make the difference between speaking to a decision-maker who interprets policy one way and speaking to one who interprets it very differently, even though both interpretations may be right in a formal sense. In this sense, it has the objective results of reducing errors and increasing efficiency which is good for academic staff, students, and universities. Mentors also help introduce new faculty members to others, invite them to social functions, and impart advice without the baggage of creating status differences. In this sense, arranged mentoring helps reduce subjective barriers to entry.

Mentoring should be recognized as an important part of the service to the department and university, and should be recognized with remuneration, reduced committee work, or flexible teaching schedules. At one university we are familiar with, arranged mentoring has been developed to a sophisticated level on these dimensions. For example, mentors are rewarded in direct proportion to the success of those they mentor – creating an added incentive in mentors to produce productive collective results. Systems like this have shown their success in the private sectors, particularly in law and consulting fields which are driven by similar systems of teamwork, interdependence, and networking.

Luce professorships and the National Science Foundations Program of visiting professorships for women provide individual permanent and temporary positions but no program is yet available of the magnitude to create a critical mass by itself. However, an internal university commitment can provide the necessary scale of resources for achieving a critical mass at least in some departments, as in the molecular biology department that we studied. A more radical suggestion, given the success of women's colleges in encouraging women's participation in science at the undergraduate level, is the development of graduate departments at some of these same institutions (Lazarus and Nair, undated).

Such a bold step would provide a place for faculty members to set an example of women organizing research groups that function collegially, effectively and differently than the male model. Serious consideration of such a course of action might lead existing graduate programs to re-evaluate their treatment of women since the resources to initiate this reform of the academic system could well be drawn from the National Science Foundation, National Institutes of Health and other agencies that support existing graduate programs.

POLICY RECOMMENDATIONS FOR DEPARTMENTS

Although culture is generally believed to be highly resistant to change, our findings suggest a few key points of intervention. Specific steps could be taken to mitigate the effects of the male scientific ethos on the recruitment of women to science and engineering. The rigidity of the existing academic structure and misperceptions of women scientists among male academics constitute formidable barriers to the entry and retention of women at the highest levels of academic science. However, the fact that qualified women who would be interested in academic research careers are now in industry or teaching colleges suggests that, should these final barriers be lowered or removed, women scientists who already exist might pursue careers at the highest levels of academic science.

Women scientists wish to legitimate an alternative model that will open up science to women's full participation. They raise equity issues in the face of strongly held beliefs on the part of many male scientists that the existing system produces enough female scientists. Because parochial ways of conceptualizing, investigating, and organizing the conduct of science have been accepted, significant sectors of the population have been excluded from full participation, and alternative perspectives and organizational styles have been repressed. As we become aware of such factors as masculine models of gender as the basis for many modes of doing science, a policy space is opened up where change can take place. Social movements and support groups organized by excluded groups, changes in departmental practices and university policies taken at the initiative of faculty and administrators, and governmental affirmative action policies and funding programs are all part of the emerging picture of science open to all talent in fact as well as by precept.

CONCLUSION: SCIENCE POLICY FOR WOMEN IN SCIENCE

Questions of gender and science have come into the foreground in sociological theory, feminist research and human resource policy

(Abir-Am, 1989; 1991). The sociology of science is moving beyond comparing men and women scientists according to implicitly masculine criteria, which have themselves come into question. Hyper-competitiveness has been attacked as counterproductive to 'good science' (Marvis, 1993), leading to premature publication and possibly to 'fudging' and fraud. It has also given us the 'smallest publishable unit', the practice of subdividing findings into numerous articles. Harvard University has recognized the problem, limiting the number of articles that can be submitted for tenure review. The definition of research achievement in terms of number of publications, with article counts accepted as a primary indicator of productivity and achievement, is ambiguous. Women publish less frequently than men but their publications are more frequently cited (Long, 1990). This finding suggests different gender styles of scientific work, with women taking a more measured approach to research. Women appear to work more intensively on a subject before making their work public.

Institutional reform that takes into account the needs of female scientists also benefits men by introducing some flexibility into career paths. How can the phenomenon that graduate departments are more active in organizing programs for undergraduates and high school students than for their own students be explained? One answer is that instead of changing a structure that the people in power are satisfied with, it is much easier to deal with other people's problems, elsewhere. Perhaps this is why there are very few programs at the graduate level even in departments that are active in organizing programs lower down the academic ladder.

Departmental change to advance the interests of women is unlikely to be widespread unless there is intervention from above, either from the leadership in the university or from the funding agencies that support the research system on which Ph.D. programs are based.

Innovative university-wide policies can change department policies; for example one might be a department review for diversity issues. Just as a U.S. university department has an external review every three years that is meant to assess the department's research, internal dynamics, and links to the wider university, it should have a separate diversity review (as in private industry), or a diversity assessment should be made part of the conventional three-year academic review. The outside referees (usually from comparable departments elsewhere) should evaluate how well the department is doing on issues of recruiting, hiring, and retaining women. Just as the review evaluates a department's coverage of subspecialty areas in the field (e.g., gene splicing in biology departments), it can assess a department's track record and current gender balance.

This change could have the effect of bringing the systematic disadvantages of women to the attention of university administrators, who could then dedicate funds to improving these conditions. It also opens up channels of discussion among female and male faculty members about the subject, which can help inform women faculty about how to choose and locate progressive departments. Finally, because these reviews are sometimes used in ranking departments, they tie department status to affirmative hiring of women – reversing some of the present and reprehensible biases. Related to the idea of an academic review with a diversity component or simply a separate diversity review, diversity committees should be set up across university departments. These committees should have the objective of assessing issues of critical mass, recruitment, hiring, and retention, identifying and diffusing models of success, and providing counsel on issues of bias.

At present, most universities only have avenues of recourse for women who experience bias or harassment that affects pay (or promotions – easy to assert but difficult to prove). These programs are necessary, but lack attention to the processes and experiences of women that shape job satisfaction and feelings of empowerment that are needed to sustain a cutting-edge research program. These committees should be staffed by men and women who are interested and motivated to effect change. Finally, a network of these committees should be created across campus in order to increase knowledge transfer about best practices and to promote the sharing of resources (e.g., to share the costs of instituting talks and seminars by elite women scientists). Members of the committee should be rewarded for their participation in annual salary and promotion reviews, with reduced teaching loads, or guaranteed teaching time slots (to help accommodate the balance between work and private life).

When departments become more inclusive, there will be less need for intervention. In the interim, programs have an important role to play in strengthening nascent female networks. Opening up existing networks to women is especially important because of the general lack of formal rules and the resistance to increased levels of formalization in academic departments. Systemic change in networks creates a level playing field for networking among both women and men. 'Normative' change, in which values informally shift over generations, with younger male and female mentors sharing non-sexist values, is a slower mechanism for the distribution of social capital to all, regardless of gender.

Critical mass was expected to be achieved through affirmative action, to clear up blockages in the pipeline on the premise that attracting a sufficient number of persons, from a previously excluded social category, will foster inclusion of others from that background. The paradox of critical mass, and the interest of many female scientists in creating an alternative mode of doing science, suggest that this is not the case. Encouraging more women to enter the pipeline is fruitless if so few emerge as professional scientists. In the face of exclusionary practices, both explicit and implicit, built into the research university system, many women Ph.D.s, see the writing on the wall and, seeking to balance work and personal life, seek employment in industry and teaching colleges. As our observations emphasize, the pipeline, a 'supply side' approach, needs to be supplemented by a focus on changing the institutional structures where science takes place.

In contrast to little more than ten years ago when we began our research, women scientists are increasingly more able to identify and reveal their sometimes painful and frequently confusing experience as graduate students, post-docs and junior faculty. Some middle-aged women retrospectively understand their personal sacrifices, adaptive strategies and defenses as responses to tenuous and hostile situations. Moreover, some male scientists now not only acknowledge the presence of bias, but openly attempt to provide women students and colleagues with strategies and support for success. Ironically, the endeavors of some male scientists to reduce marginalization shows that real inequities actually do exist. As mentors they provide the primary relationship required by every young scientist to learn the craft, the unwritten rules, and means of entry into social networks crucial for continued growth. Thus our findings show that in roles of power and authority, both male and female scientists are able to re-create departments into genuinely democratic institutional contexts. However, they are few and far between.

In response to relying on 'critical mass' as the panacea for change, we argue that only in democratic departments does the notion of critical mass really work. We have found that 'critical mass' is meaningless when women are isolated and unknown to each other, when affiliation with other women is too stigmatizing, or the female faculty model available reflects an archaic, male stereotype impossible to emulate or incorporate into a contemporary professional identity. True critical mass occurs because an informal grapevine attracts more women students and staff who are then integrated into the department as a whole. Rather than a simple statistic of 15% or more, a number that we have found frequently and erroneously reflects a large proportion of foreign students rather than American women (erroneously, because if 90% of that 15% are foreign students who remain in their own subculture, women of each nationality may be very isolated), the true power in the number reflects a department that is cohesive, inclusive and not isolating. Without the anxiety of exclusion and lowered status, women faculty members are not as inhibited in acting on behalf of female students and, therefore, are able to serve as authentic role models. Moreover, their energy is not as depleted by defensive operations around tenure, the burden of tokenism and apprehension of interrupting their careers to have children.

A key factor in overcoming the problems posed by the paradox must be university-wide policies on child-care, parental leave, recruitment and retention, and slowing of the tenure clock. At the departmental level, junior faculty who assume responsibilities as mentors and role models functions should be credited in tenure reviews. Tokenism must be eschewed: many departments aggressively court a few female stars while most women languish in continued discrimination.

Nevertheless, the ability of departments to defend traditional academic practices as 'gender neutral' should not be underestimated nor should willingness to reform be overestimated. For departments unable to reform themselves, outside pressures provide the necessary incentive. A representative of WISE (Women in Science and Engineering at Columbia University) recently suggested that the National Science Foundation cut off grants to universities without a minimum number of female faculty members in science and engineering departments. Indeed, NSF has mandated that absence of women at conferences that it funds will be taken as prima facie evidence of discrimination.

Law suits to redress discriminatory acts are expensive and time-consuming, even when successful. Until quite recently, courts were generally unwilling to review academic decisions on substantive grounds; only matters of procedure were typically subject to judicial review. Gender discrimination has now been accepted as a valid basis for law suits challenging academic decisions, following widespread acceptance of its legitimacy in other workplaces. Jenny Harrison, a mathematician at the University of California, Berkeley, was recently granted tenure after such a suit. The recognition she received for a series of significant results made the initial negative decision a matter of some embarassment to the mathematical community. Special dispensation for academic institutions, whether in the courts or Congress Legislatures, is disappearing as universities are held to ethical, legal, and financial standards common to all public institutions.

Participation of all groups in society is a basis for the public support of science. The legitimation of science, the moral injunction to achieve equity and the strategic interest of each nation in utilizing talent to its fullest extent are reasons for change. In the U.S., Neal Lane, the former director of NSF, has called upon the research community to act in its own interest and make a conscious effort 'to integrate itself into the larger community' by more closely reflecting the demographic composition of the population. Equal representation of women and men in scientific professions would counter the elitist image of science and hopefully earn increased support for allocation of public resources to science.

The most important change required is a broader, more flexible, model of the relationship between work and personal life to make scientific as well as other highly demanding professions equally open to all persons of talent, irrespective of gender. The relationship of scientific work to other spheres of life needs to be rethought, to make it more compatible not only with female aspirations and socialization but with emerging male wishes to find a better balance between career and personal life. As Alice Rossi perceptively put it, 'marriage, parenthood and meaningful work are major experiences in the adventure of life. No society can consider that the disadvantages of women have been overcome so long as the pursuit of a career exacts a personal deprivation of marriage and parenthood, or the pursuit of happiness in marriage and family life robs a woman of fulfillment in meaningful work.' How can this goal be achieved?

BEYOND POLICY INTERVENTIONS
Policy change cannot affect inherent attitudes and prejudices. Change of that nature appears to emanate from those in power within the department. They become the role model for the role models. In order for women to cope with the vicissitudes of gender discrimination, they require the armament of the reality about the paradoxes within the culture of science. We believe that prior to entering graduate school young women need the opportunity to interact informally with senior graduate students and faculty members who can talk candidly about

these issues. Workshops could be organized to provide knowledge of the unwritten rules and the strategies required to thrive, as well as a first-hand experience of the interpersonal connections and acceptance that are possible among scientists. In contrast to repeated concerns that bringing the difficulties for women in science out in the open will only dissuade them from pursuing science, we suggest the opposite.

Making potential problems more immediately recognizable, and solutions and strategies more attainable, is empowering and inhibits the debilitating process that all too frequently occurs only after the student is in a graduate program and has no means to identify what she is experiencing. That is when the enervating feelings of anxiety, shame and self-blame begin. We also believe that venues for bringing these issues out in the open are as imperative for upcoming junior faculty members as they are for graduate students. Not to speak forthrightly is only a re-enactment of the denial that occurs in those departments in which marginalization endures. In such departments rationalizations are pervasive and change does not occur. Most of all, if it is true as it has been suggested that women really do need to respond better to negative 'kicks' (Sonnert and Holton, 1996), they will first have to know what those kicks may be.

Our research strongly indicates that until the social context alters, women will have to understand the critical role of their advisor before the choice is ever made. Women need to know those personal and professional characteristics of this uniquely important individual who will either enhance or diminish their chances of attaining goals and developing a professional identity. This is crucial since a professional identity is inextricably linked to a 'social identity' in which the esteem of others provides recognition and serves to enhance self-esteem (Berger, 1967).

In contrast to earlier assumptions in which all women academics were perceived and touted as automatic role models and mentors, young students need to understand and be able to identify the attributes necessary in a good mentor of either sex. In this respect, the notion of 'gender differences' is again called into question: some male advisors can be excellent mentors for all of their students regardless of sex. We believe that there is evidence that adherence to cultural prescriptions and proscriptions around notions of gender depends on the individual and, therefore, is not immutable.

Rather than rely solely on quantitative analysis which frequently masks what is really going on in people's lives, we have attempted to understand the real experiences of women scientists as they have shared their understanding of their experiences. We believe that multiple paradoxes abound, potentially double-binding women at every juncture within the pipeline. It begins with marginalization and isolation and the demand for autonomous, independent functioning within an activity which is, for men, highly social and socializing. It is exacerbated when adaptive attempts for affiliation through women's groups are labeled as indicating 'special needs'. It is compounded when similarly isolated women faculty members are offered up as a solution to institutional problems.

It needs to be recognized that some of the solutions and mechanisms employed by successful women scientists of an older generation are no longer relevant in a different historical and social context. Until these and many other paradoxes for women in science can be freely examined and articulated without the subtle threat that an inherent and innate 'difference' will be exposed, the contradiction between what too many in the scientific community choose to believe exists and what actually occurs will continue to impede the growth and development of too many gifted women.

The late Betty Vetter, founder of the Commission on Professionals in Science and Technology, pointed out the waste engendered by persisting barriers that cut short scientific careers in which there has been considerable personal and public investment. It is high time to remove the barriers that impede women and minorities from successfully negotiating the 'critical transitions' of scientific and engineering education and career. Norms of science that incorporate both traditional male and female perspectives into a broader non-sexist framework would free both experimentation and verification of knowledge from the exclusionary oppositions in which feminine is automatically conceived of as antithetical to 'good science' (Keller, 1980).

Under these conditions, with impersonal evaluation a component of the social structure of science, Maria Mitchell's exhortation (quoted in Enna, 1993) would become a reality: ' . . . no woman should say, "I am but a woman!" But a woman! What more could you ask to be?'

Appendix

SURVEY METHODOLOGY

Our survey was mailed to the faculty in the departments of biology, biochemistry, chemistry, computer science, engineering, and physics at a large private Midwestern University in May of 1997. Emeritus, adjunct, and visiting faculty members were not contacted because they are normally peripheral to department activity. Data collection was completed by September 1997. The survey was diskette-based and computer-administered and took approximately 30 minutes to complete. The electronic survey guided respondents through the items (with appropriate branching) and created a data file on the diskette, which was returned to us. This procedure reduced errors of data entry by respondents and data entry assistants, and ensured that all respondents answered all items. All faculty members were sent a cover letter that briefly described the purpose of the study, assured confidentiality, gave references to our previous work in scientific journals of the hard sciences and at the National Science Foundation, and described the instructions for the use of the survey. Included along with the cover letter were a diskette and a pre-addressed envelope for returning the diskette. Table A1 describes the characteristics of our sample.

Faculty members who did not respond in three weeks or who used Macintosh computers were asked to set up a time for a telephone interview by a trained research assistant who entered responses in real time into a PC. Unfortunately, the biology department used Macintosh computers exclusively, which lowered the response rate of this department. Table A1 displays the response rate, which averaged 56% and ranged from 38% to 76%. This response rate is similar to other network surveys of organizations (Podolny and Baron, 1997; Burt 1992; Granovetter, 1973).

Table A1: *Sample characteristics and response rate*

Department	Number of Faculty			Number of Respondents		
	Men	Women	% Women	Men	Women	Response rate
Biology	28	6	21.4	12	1	38 %
Biochemistry	23	13	57	14	9	63 %
Chemistry	25	2	8	17	2	76 %
Computer Science	8	0	0	4	0	50 %
Engineering	33	5	15	22	1	57 %
Physics	27	3	11	16	0	53 %
Total	144	29	20	84	13	

We explored whether there was non-response bias by examining differences in the demographic background (age, gender, rank, and income) of faculty members who responded immediately and those who were interviewed by phone. There appeared to be few differences across these categories using O'Brien's method (Stata, 1996). There also did not appear to be a difference between responders and non-responders in terms of prestige of Ph.D.-granting institution (most had received degrees from prestigious schools), although full professors were more likely to respond after the first mailing. Lastly, the feedback we received from faculty members who could not participate indicated that it was due to randomly distributed factors (deadlines, annual meetings, grants, travel, etc.).

REGRESSION ANALYSIS

We used an ordered logit model to estimate the effect of social capital on the rate of publication. This model estimates the relationship between a categorical and an ordered dependent variable – 'no output,' 'low output,' 'medium output,' and 'high output' – and a set of independent variables (Greene, 1993). An ordinal variable is a variable that is categorical and ordered, for instance, 'no output,' 'low output,' 'medium output,' and 'high output.' In an ordered logit, an underlying probability score of how a one unit change in an independent variable affects the change in probability of intensity of output is estimated as a linear function of the independent variables and set of cut points. The probability of observing outcome i corresponds to the probability that the estimated linear function, plus random error, is within the range of the cut points estimated for the outcome:

$$\Pr(\text{outcome } j = i) = \Pr(K_{i-1} < B_1 x_j + \ldots + B_k x_{kj} + u_j \leq K_i)$$

One estimates the coefficients $B_1, B_2, \ldots, B_k$ along with the cut points $K_1, K_2, \ldots, K_{I-1}$, where I is the number of possible outcomes. All of this is a generalization of the ordinary two-outcome logit model. The ordered logit predictions are then the probability that outcome $j + u_j$ lies between a pair of cut points K_{i-1} and K_i (Stata 1996).

The attractive modeling feature of the ordered logit is that the substantive numerical values of the dependent variable are unimportant (Greene, 1993). In ordinary regression, arbitrarily assigning the number values of 4 to the 'high output' category, 3 to the 'medium output' category, 2 to the 'low output' category, and so on is inappropriate because different numeric values (say 10 versus 8 for 'high output') would obtain different estimates. This is not true in an ordered logit model. All that is necessary is that larger numbers correspond to more intense outcomes or levels of usage.

DEPENDENT VARIABLE

The criterion variable, research productivity, was measured by asking respondents to indicate their number of publications using the following interval scale: (1) no publications, (2) 1–2, (3) 3–4, (4) 5–10, (5) 11–20, (6) 21–50, and (7) more than 50 publications. This question was repeated for (a) articles published, (b) book chapters published, and (c) books published, in order to capture the range of publications. The value for each respondent on (a), (b), and (c) was summed and divided by three to derive an average productivity score. A simple sum of (a) to (c) produced similar results to those reported in terms of significance and size of coefficients. The seven preceding categories were united into four categories by joining adjacent categories and dropping the first two categories, which had no observations. The use of a four-category dependent variable made the results more intuitive and produced results that were similar to the more complex eight-category model. Table A2 displays the distribution of the dependent variable for all faculty and untenured faculty. The numerical value '0' corresponds to the low output category (Less than 4), '1' corresponds to the low-to-medium category (5 to 10), '2' corresponds to the medium-to-high category (11 to 20), and '3' corresponds to the high output category (more than 21). Table 12.4 displays three cut points because one of the four categories is a reference category, which in this analysis corresponds naturally to category 0 (see Table A2).

Table A2: *Distribution of research productivity of faculty*

Number of publications reported	All faculty		Untenured faculty	
	Men	Women	Men	Women
Less than 4	3%	7%	7%	10%
5 to 10	12%	43%	20%	45%
11 to 20	28%	25%	45%	25%
More than 21	57%	25%	28%	20%

INDEPENDENT VARIABLES

Token Overload, Power Imbalance, Number of Strong Ties, and Number of Bridge Ties were measured using the items and scales described in Chapter 12. We squared the number of strong ties to examine our hypothesis that an intermediate level of strong ties is positively associated with research productivity. Number of Co-authors was simply the number reported by the respondent.

CONTROL VARIABLES

Following prior research, we controlled for human capital and demographic factors with the following measures (Seashore *et al.*, 1989; Cole, 1992). We created an indicator variable, Gender, which was coded 0 for female and 1 for male, and Tenured which was coded 1 for tenured and 0 for non-tenured. Professional Age measured number of years since Ph.D. and Age in Years measured age of respondent. Cole (1992) reported that Professional Age has been found to have a statistically significant but small positive effect on getting grants. Age in Years has similarly been the center of many studies of scientific productivity. Cole (1992) noted that scientific creativity is commonly believed to decline with age, but noted that the empirical evidence is mixed. Research Budget Level controls for the level of the faculty's research budget (in dollars divided by 100), a factor positively associated with research productivity (Seashore *et al.*, 1989; Cole, 1992). Finally, we created an indicator variable called Post-Doc (1=Yes) to control for the positive effect of a post-doctorate fellowship on research productivity (Long and McGinnis, 1981).

Two variables were added to control for the synchronicity problem, which is the condition that the number of publications reported was probably affected by ties that existed prior to the ties reported at the time of the survey. One way to deal with this problem is to assess the historical stability of the network over the period of publishing reported. Network Turnover measures the level of turnover in an individual's network over the past two years. Another control variable that attempts to mitigate the synchronicity problem is Average Tie Duration, a variable that measures the average time the respondent has known the contacts named. The assumption underlying these controls is that contacts that have endured more than three years are likely to be long-term ties and in existence over the course of the reported number of publications.

Bibliography

Abir-Am, Pnina. 1989. 'Science Policy or Social Policy for Women in Science: Lessons From Historical Case Studies.' Paper presented at NATO Advanced Studies Institute In Science Policy, Il Cioccio, Italy, October 1–13.

Abir-Am, Pnina. 1991. 'Science Policy for Women in Science: from historical case studies to an agenda for women in science.' History of Science Meetings, Madison, Wisconsin, 2 November.

Abir-Am, Pnina and Dorinda Outram. 1986. *Uneasy Careers and Intimate Lives.* New Brunswick: Rutgers University Press.

Abir-Am, Pnina and Dorinda Outram, eds. 1987. *Uneasy Careers and Intimate Lives, Women in Science 1789–1979.* New Brunswick: Rutgers University Press.

Acar, Feride. 1991. 'Women in Academic Science Careers in Turkey.' *Women in Science: Token Women or Gender Equality.* In Stolte-Heiskanen, Veronica, *et al.*, eds. Oxford: Berg Publishers.

Adelman, Clifford. 1998. *Women and Men of the Engineering Path: A Model for the Analyses of Undergradute Careers.* Washington DC: U.S. Government Printing Office.

Alemany, Maria Carime. 1991. 'Is to be an Engineer still a Masculine Career in Spain? Notes on an Ambiguous Change in University Technical Education.' pp. 215–26. In *Women in Science: Token Women or Gender Equality.* Stolte-Heiskanen, Veronica, *et al.*, eds. Oxford: Berg Publishers.

Alexander, V.D. and P.A. Thoits. 1985. 'Token Achievement.' *Social Forces* 64: 332–40.

American Association of University Women. 1990. '*Shortchanging Girls, Shortchanging America.*' Washington DC.

American Institute of Physics (AIP). 1988. *Physics in the High Schools.* New York: AIP (#R-340).

American Institute of Physics (AIP). 1991. *Enrollments and Degrees.* New York: AIP (#R$-151.28).

Ananieva, Nora. 1991. 'Women and Science in Bulgaria: the long hurdle races.' In *Women in Science: Token Women or Gender Equality?* Veronica Stolte-Heiskanen, ed. Oxford: Berg Publishers.

Angier, Natalie. 1991. *World's Women: Trends and Statistics 1970–1990.* (June 1991) United Nations Pubns.

Angier, Natalie. 1995. 'Why Science Loses Women in the Ranks.' *New York Times* Sunday, May 14: 4/5.

Barber, Leslie A. 1995. 'U.S. Women in Science and Technology, 1960-1990.' *Journal of Higher Education* 66(2) (March/April 1995): 213–34.

Barinaga, Marcia. 1993. Is There a Female Style in Science?' *Science* 260: 384–91.

Barnett, R. 1974. 'Sex Differences and Age Trends in Occupational Reference and Occupational Prestige.' *Journal of Counseling Psychology* 22: 35–8.

Baron, James N. 1984. 'Organizational Perspectives on Stratification.' *Annual Review of Sociology* 10: 37–69.

Baron, James N. and William T. Bielby, 1980. 'Bringing the Firms Back In: Stratification, Segmentation, and the Organization of Work.' *American Sociological Review* 45: 737–65.

Bem, Sandra Lapses. 1983. 'Gender Schema Theory and Its Implications for Child Development: Raising Gender-aschematic Children in a Gender-schematic Society.' *Journal of Women in Culture and Society* **8**(4): 598–616.

Benbow, Camilla P. and Julian C. Stanley. 1980. 'Sex Differences in Mathematical Ability: Factor or Artifact?' *Science* **220**: 1262–64.

Benjamin, Marina. 1991. *Science and Sensibility: Gender and Scientific Inquiry; 1780–1945.* London: Basil Blackwell.

Berger, Peter. 1967. 'Some General Observations on the Problem of Work.' *The Human Shape of Work*. Peter Berger, ed. New York: The MacMillan Company.

Berryman, Sue E. 1983. *Who Will Do Science? Women and Minorities in Science; Trends and Causes.* Report to the Rockefeller Foundation, NY.

Bian, Yanjie. 1997. 'Indirect Ties, Network Bridges, and Job Searches in China.' *American Sociological Review* **62**: 366–85.

Birns, B. 1976. 'The Emergence and Socialization of Sex Differences in the Earliest Years.' *Merrill-Palmer Quarterly* **22** (1976): 229–52.

Blagojevic, Marina. 1991. 'Double-Faced Marginalisation: Women in Science in Yugoslavia.' pp. 75–93. In *Women in Science: Token Women or Gender Equality*. Stolte-Heiskanen, Veronica, *et al.* eds. Oxford: Berg Publishers.

Blazquez, Norma Graf. 1991. 'The Role of Women in the Developmen of Science and Technology in Mexico.' Proceedings of the 91CWES, University of Warwick, UK, p. 15–18D.

Block, Jeanne H. 1984. *Sex Role, Identity and Ego Development*. San Francisco, CA: Jossey-Bass Publishers.

Boas, Franz. 1903. *The Kwaikutl Indians of Vancouver Island*. New York: Bril.

Bourdieu, Pierre and Loic J. D. Wacquant. 1992. *An Invitation to Reflexive Sociology*. Chicago: University of Chicago Press.

Burt, Ronald S. 1992. *Structural Holes: The Social Structure of Competition*. Boston, MA: Harvard University Press.

Burt, Ronald. 1997. 'The Contingent Value of Social Capital.' *Administrative Science Quarterly* **42**: 339–65.

Burt, Ronald S. 1998. 'The Gender of Social Capital.' In *Solidarity and Inequality*. Siegwart Lindenberg, Wout Ultee, and Rie Bosman, eds. Forthcoming.

Bush, Vannevar. 1945. *Science – The Endless Frontier*. Washington DC: National Science Foundation.

Cacoullos, Ann. 1991. 'Women, Science and Politics in Greece' pp. 135–146. In *Women In Science: Token Women or Gender Equality*. Stolte-Heiskanen, Veronica, *et al.*, eds. Oxford: Berg Publishers.

Campbell, George Jr, Ronni Denes and Catherine Morrison. 2000. *Access Denied: Race, Ethnicity, and the Scientific Enterprise*. New York: Oxford University Press.

Carrasco, Cecilia Lopez. 1995. 'Nueva Disciplina, Viejos Prejucios?' Paper presented at V Colloquio de Estudios de Genaro, organized by the Programa Universitario de Estudios de Genaro (PUEG), UNAM, 16–20 October 1995.

Carson, Nandy and Darryl Chubin. 1992. 'Women in Science and Engineering: A Data Update.' U.S. Congress OTA Seminar, May 25.

Catsambis, Sophia. 1994. 'The Path to Math: Gender and Racial-Ethnic Differences in Mathematics Participation from Middle School to High School.' *Sociology of Education* **67**(3): 199–215.

Catsambis, Sophia. 1995. 'Gender, Race-Ethnicity and Science Education in the Middle Grades.' *Journal of Research in Science Teaching* **32**: 243–57.

Cattell, R.B. and J.E. Drevdahl. 1973. 'A Comparison of the Personality Profile (16 P.F.) of Eminent Researchers and That of Eminent Teachers and Administrators, and of the

General Population' in *Science As A Career Choice: Theoretical and Empirical Studies*. Eiduson, Bernice T. and Beckman, Linda J., eds. New York: Russell Sage Foundation.

Chestang, Leon. 1982. 'Work, Personal Change, and Human Development.' In *Work, Workers and Work Organizations: A View From Social Work*. Sheila Akabas and Paul A. Kurzman, eds. Englewood Cliffs, NJ: Prentice Hall.

Chodorow, Nancy. 1978. *The Reproduction of Mothering: Psychoanalysis and the Sociology of Gender*. Berkeley: University of California Press.

Chodorow, Nancy. 1990. *Gender, Relation, and Difference in Psychoanalytic Perspective*. New York: New York University Press.

Cole, Jonathan and Harriet Zuckermann. 1987. 'Marriage and Motherhood and Research Performance in Science.' *Scientific American*, no. 256.

Cole, Stephen. 1992. *Making Science Between Nature and Society*. Cambridge, MA: Harvard University Press.

Coleman, James S. 1988. 'Social Capital in the Creation of Human Capital.' *American Journal of Sociology* **94**: S95–S120.

Coleman, James S. 1990. *Foundations of Social Theory*. Boston: Harvard University Press.

Condry, John. 1984. 'Gender Identity and Social Competence.' *Sex Roles: A Journal of Research*, **II** (No. 5/6): 484–507.

Couture-Cherki, Monique. 1976. 'Women In Physics.' In *Ideology of/in the Natural Sciences*. Hilary Rose and Steven Rose, eds. Cambridge: Schenkman.

Curtin, Jean M., Geneva Blake and Christine Cassagnau. 1997. 'The Climate for Women Graduate Students in Physics.' *Journal of Women and Minorities in Science and Engineering* **3**, 95–117.

Davis-Blake, Alison and Brian Uzzi. 1993. 'Determinants of Employment Externalization: A Study of Temporary Workers and Independent Contractors.' *Administrative Science Quarterly* **38**: 195–223.

Dickson, David. 1984. *The New Politics of Science*. New York: Pantheon.

Didion, Catherine, 1993. 'Attracting Graduate and Undergraduate Women as Science Majors.' *Journal of College Science Teaching*.

Dimant, Stephanie. 1995. 'Science is for Childless Women.' *New York Times* Wednesday, May 17.

Dresselhaus, Mildred, G. Dresshaus and P.C. Eklund. 1996. *Science of Fullerenes and Carbon Nanotubes*. Academic Press.

Drucker, Philip and Robert Heizer. 1967. *To Make My Name Good: A Reexamination of the Southern Kwakiutl Potlatch*. Berkeley: University of California Press

Dupree, Andrea. 1991. Interview by Jonathan Cole and Harriet Zuckerman in *The Outer Circle: Women in the Scientific Community*. Zuckerman, Harriet, Jonathan Cole and John T. Breuer, eds. New York: Norton.

Eccles, Jacqueline and Janis E. Jacobs. 1986. 'Social Forces Shape Math Attitudes and Performance.' *Journal of Women in Culture and Society*. **II**(21): 367–89.

Eiduson, Bernice. 1973. 'The Scientists' Personalities.' In *Science As A Career Choice: Theoretical and Empirical Studies*. 1973. Bernice Eiduson and Linda Beckman, eds. New York: Russell Sage Foundation.

Emmett, Arielle. 1992. 'A Woman's Institute of Technology.' *Technology Review* (April) pp. 16–18.

Enna, Anne. 1993. 'In Mitchell's 19th-Century Vision, Women Find New Dreams.' *Nantucket Inquirer and Mirror*, August 26, DI.

Epstein, Cynthia. 1970. *Woman's Place: Options and Limits in Professional Careers*. Berkeley: University of California Press.

Etzkowitz, Henry. 1971. 'The Male Nurse: Sexual Separation of Labor in Society.' *Journal of Marriage and the Family*, August 1971.

Etzkowitz, Henry. 1991. 'Individual Investigators and their Research Groups.' *Minerva*, Spring 1992.

Etzkowitz, Henry. 1992. 'Individual Investigators and Their Research Groups.' *Minerva* (Spring).

Etzkowitz, Henry and Lois Peters. 1991. 'Profiting from Knowledge: Organizational Innovations and the Evolution of Academic Norms' *Minerva* **29** (Summer) pp. 133–66.

Etzkowitz, Henry, Carol Kemelgor, Michael Neuschatz, Brian Uzzi, and Joseph Alonzo. 1994. 'The Paradox of Critical Mass for Women in Science.' *Science* **266**: 51–3.

Etzkowitz, Henry and Loet Leydesdorff, eds. 1997. *Universities in the Global Knowledge Economy: A Triple Helix of University–Industry–Government Relations*. London: Cassell.

Etzkowitz, Henry and Carol Kemelgor. 1998. The Role of Research Centres in the Collectivisation of Academic Science. *Minerva* **36** (No. 2, Autumn) pp. 271–88.

Etzkowitz, Henry. In Press. *The Second Academic Revolution: M.I.T. and the Rise of Entrepreneurial Science*. London: Gordon and Breach.

Fagot, B. 1978. 'Sex-Determined Consequences of Different Play Styles in Early Childhood.' Paper presented at the Annual Meeting of the American Psychological Association, Toronto, Canada, August 1978.

Female Graduate Students and Research Staff in the Laboratory for Computer Science and the Artifical Intelligence Laboratory at M.I.T., February, 'Barriers to Equality in Academia: Women in Computer Science at M.I.T.' Unpublished Manuscript.

Fernandez, Roberto M. and Nancy Weinberg. 1997. 'Sifting and Sorting: Personal Contacts and Hiring in a Retail Bank.' *American Sociological Review* **62**: 883–902.

Firnberg, H. 1987. 'Frauen und Forschung' (Women and Research) in *Frauenstudium and academische Frauenarbeit in Osterreich* (Women's Higher Education and Academic Work in Austria) *1968–1987*. Vienna: Austrian Federation of University Women. pp. 17–29.

Fox, Mary Frank. 1989. 'Women in Higher Education: Gender Differences in the Status of Students and Scholars.' In *Women: A Feminist Perspective*. Jo Freeman (ed.) Mountain View CA. Mayfield Publishing Co.

Fox-Keller, Evelyn. 1985. *Reflections on Gender and Science*. New Haven: Yale University Press.

Fox-Keller, Evelyn. 1980. 'How Gender Matters, or, Why It's So Hard for Us to Count Past Two.' In *Perspectives on Gender and Science*. Harding, Jan., ed., 1986. London: Falmer Press.

Fox-Keller, Evelyn. 1983. *A Feeling for the Organism: The Life and Work of Barbara McClintock*. San Francisco: W. H. Freeman.

Friedkin, Noah E. 1980. 'A Test of the Structural Features of Granovetter's 'Strength of Weak Ties' Theory.' *Social Networks* **2**:411–22.

Friedman, Gloria and Barry Protter. 1995. 'Gender and Sexuality'. In *Handbook of Interpersonal Psychoanalysis*. Marylou Lionells, John Fiscalini, Carola Mann and Donnel B. Stern, eds. Hillsdale NJ: The Analytic Press.

Gabor, Andrea. 1995. *Einstein's Wife: Work and Marriage in the Lives of Five Great Twentieth-Century Women*. New York: Viking.

Gaudart, Dorothea. 1991. 'The Emergence of Women into Research and Development in the Austrian Context' pp. 9–34. In *Women in Science: Token Women or Gender Equality*. Stolte-Heiskanen, Veronica, *et al.*, eds. Oxford: Berg Publishers.

Gerson, Kathleen. 1985. *Hard Choices: How Women Decide about Work, Career and Motherhood*. Berkeley: University of California Press.

Gibbons, Ann. 1993. 'Women in Science: Pieces of the Puzzle.' *Science* **255**: 1368.

Gilligan, Carol and Lynn Brown. 1990. *Meeting at the Crossroads: The Psychology of Women and Girls' Development.* Cambridge: Harvard University Press.

Goldberg, Carey. March 23, 1999. *The New York Times.* M.I.T. Acknowledges Bias Against Female Professors.

Gould, L. W. 1972. 'A Fabulous Child's Story.' *Ms.* December, 1972: 74–76.

Granovetter, Mark. 1973. 'The Strength of Weak Ties.' *American Journal of Sociology* 78(6): 1360–80.

Granovetter, Mark. 1985. 'Economic Action and Social Structure: The Problem of Embeddedness.' *American Journal of Sociology* 91: 481–510.

Greene, William H. 1993. *Econometric Analysis.* New York: MacMillan Publishing Company.

Gutek, B.A. 1985. *Sex and the Workplace.* San Francisco: Jossey Bass.

Haas, Violet and Carolyn Perucci. 1986. *Women in the Scientific and Engineering Professions.* Ann Arbor: University of Michigan Press.

Hare-Mustin, Rachel T. and Jeanne Maracek. 1988. 'The Meaning of Difference: Gender Theory, Postmodernism and Psychology.' *American Psychologist.* June 1988.

Haritos, Rosa and Ronald Glassman. 1990. 'Emile Durkheim and the Sociological Enterprise ' In The Renascence of Sociological Theory: Classical and Contemporary. Henry Etzkowitz and Ronald Glassman (eds.) 1991. Itasca, Illinois: Peacock Press, 1991.

Healey, Peter. 1992. Personal communication to Henry Etzkowitz.

Hicks, Esther. 1991. 'Women at the Top in Science and Technology fields. Profile of Women Academics at Dutch Universities.' In *Women in Science: Token Women or Gender Equality.* Stolte-Heiskanen, Veronica, *et al.*, eds. Oxford: Berg Publishers.

Hirschman, Albert O. 1970. *Exit, Voice and Loyalty.* Cambridge, MA: Harvard University Press.

HMSO. 1994. *The Rising Tide: A Report on Women in Science, Engineering and Technology.* London: Her Majesty's Stationary Office

Hoffman, L. W. 1977. 'Changes in Family Roles, Socialization and Sex Differences.' *American Psychologist.* 32: 644–57.

Hollenshead, Carol, Stacy Wenzel, Barbara Lazarus, and Indira Nair. 1994. 'Influences on Women Graduate Students in Engineering and Sciences: Rethinking a Gendered Institution.' Paper presented at CURIES Conference on Women in Science, Mathematics and Engineering, Wellesley College, 19–22 May.

Hornig, Lily. 1987. 'Women Graduate Students: A Literature Review and Synthesis.' In *Women: Their Underrepresentation and Career Differentials in Science and Engineering.* Linda Dix, ed. Washington DC: National Academy Press

Horning, Beth. 1993. 'The Controversial Career of Evelyn Fox-Keller.' *Technology Review* (January) pp. 59–68.

Howell, Elizabeth. Private communication to Carol Kemelgor.

Hyde, Janet Shibley. 1994. 'Can Meta-Analysis Make Feminist Transformations in Psychology?' *Psychology of Women Quarterly* 18 Cambridge University Press.

Ibarra, Hermina. 1992. 'Homophily and Differential Returns: Sex Differences in Network Structure and Access in an Advertising Firm.' *Administrative Science Quarterly* 37: 422–47.

Ibarra, Hermina and Lynn Smith-Lovin. 1997. 'New Directions in Social Network Research on Gender and Organizational Careers.' In *Handbook of Organizational Behaviour.* S. Jackson and C. Cooper, eds. Sussex, England: J. Wiley.

Ibarra, Hermina and Steven B. Andrews. 1993. 'Power, Social Influence, and Sense Making: Effects of Network Centrality and Proximity on Employee Perceptions.' *Administrative Science Quarterly* 38(2): 277–303.

Kanter, Rosabeth M. 1977. Men and Women of the Corporation. New York: Basic Books.

Kaplan, Alexandra and Janet Surrey. 1984. 'The Relational Self in Women.' In *Developmental Theory and Public Policy*. Lenore Walker, ed. Thousand Oaks, CA: Sage Publications.

Kaplan, Eugene H. 1980. 'Adolescents, Age Fifteen to Eighteen: A Psychoanalytic Developmental View.' *The Course of Life: Psychoanalytic Contributions Toward Understanding Personality Development*. Greenspan and Pollock, eds. NIH.

Kenney, Martin. 1986. *Biotechnology: the University-Industrial Complex*. New Haven, CT: Yale University Press.

Kohlberg, L. 1966. 'A Cognitive-Developmental Analysis of Children's Sex-Role Concepts and Attitudes.' In *The Development of Sex Differences*. E. E. Maccoby, ed. Stanford, CA: Stanford University Press.

Konrad. A. M. 1986. *The Impact of Workgroup Composition on Social Integration and Evaluation*. Ph.D. dissertation, Claremont Graduate School.

Koput, Kenneth W., Walter W. Powell, and Laurel Smith-Doerr. 1998. 'Interorganizational Relations and Elite Sponsorship: Mobilizing Resources in Biotechnology.' In *Corporate Social Capital*. A. J. Leenders and Shaul Gabbay, eds. New York: Addison Wesley.

Koval, Vitalina. 1991. 'Soviet Women in Science.' In *Women in Science: Token Women or Gender Equality?* Veronica Stolte-Heiskanen, ed. Oxford: Berg Publishers.

Krackhardt, David and J. R. Hanson. 1993. 'Informal Networks: The Company Behind the Chart.' *Harvard Business Review* July–Aug.: 104–9.

LaFollette, Marcel. 1988. 'Eyes on the Stars: Images of Women Scientists in Popular Magazines.' *Science, Technology and Human Values* 13 (3 & 4) (Summer and Autumn): 262–75.

Lane, Nancy *et al*. 1994. *The Rising Tide: A Report on Women in Science, Engineering and Technology*. London: HMSO.

Lazarus, Barbara and Indira Nair. Undated. 'The Case for a Women's Institue for Engineering.' Pittsbugh: Carnegie Mellon University. Unpublished manuscript.

Lear, Linda. 1997. *Rachel Carson: Witness for Nature*. New York: Henry Holt.

Lemoine, W. 1994. 'Women's Place in Science in Venezuela (1875–1958).' *Community Service Review* (UCS), nos. 35–6. pp. 55–9.

Lin, Nan, Walter Ensel, and Jan Vaughn. 1981. 'Social Resources and Strength of Ties: Structural Factors in Occupational Status Attainment.' *American Sociological Review* 46: 393–405.

Lindenberg, S., W. Atlee and R. Bosman (eds.) 1998. 'The Gender of Social Capital'. *Solidarity and Inequality* (forthcoming).

Linowitz, Sol M. and Martin Mayer (contributor). 1996. *The Betrayed Profession: Lawyering at the End of the Twentieth Century*. Reprint edition (February 1996) Johns Hopkins University Press.

Long, Scott and Robert McGinnis. 1985. 'The Effects of the Mentor on the Academic Career'. *Scientometrics* 7(3–6) pp. 255–80.

Long, Scott. 1990. 'The Origins of Sex Differences in Science.' *Social Forces*.

Longino, Helen. 1987. 'Can there be a Feminist Science?' *Hypatia* 2,3 (Fall): 51–64.

Lopez, Cecilia Farrasco. 1995. 'Women and Computing at the National Autonomous University of Mexico (UNAM): An Interpretation.' Paper presented at the Women, Gender and Science Question, University of Minnesota, May 12–14.

Lovitts, Barbara. 1996. 'Leaving the Ivory Tower: A Sociological Analysis of the Causes of Departure From Doctoral Study.' Ph.D. Dissertation, Department of Sociology. University of Maryland.

Luukkonen-Gronow, Terttu and Veronica Stolte-Heiskanen. 1983. 'Myths and Realities of Role Incompatibility of Women Scientists.' *Acta Sociologica* 26(3/4): 267–80.

Luukkonen-Gronow, Terttu. 1987. 'University Career Opportunities for Women in Finland in the 1980s.' *Acta Sociologica* **30**(2): 193–206.

Maccoby, E. and C. Jacklin. 1966. *The Psychology of Sex Differences.* Stanford, CA: Stanford University Press.

Malinowski, Bronislaw. 1922. *Argonauts of the Western Pacific.* New York: E.P. Dutton.

Marsden, Peter V. and Karen E. Campbell. 1984. 'Measuring Tie Strength.' *Social Forces* **63**(2): 482–501.

Marvis, Jeffrey. 1992. 'Radcliffe President Lambastes Competitiveness in Research.' *The Scientist.* January 20: 3.

Mason, Joan. 1991. 'The Invisible Obstacle Race.' *Nature* **353**: 205–6.

Max, Claire. 1982. 'Career Paths for Women in Physics.' In *Women and Minorities in Science: Strategies for Increasing Participation.* Sheila Humphreys, ed. Boulder: Westview.

McClelland, David C. 1973. 'On the Psychodynamics of Creative Physical Scientists.' In *Science As A Career Choice: Theoretical and Empirical Studies.* 1973. Bernice Eiduson and Linda Beckman, eds. New York: Russell Sage Foundation.

McDonough, Patricia M., Marc J. Ventresca, and Charles Outcalt. 1999 (forthcoming). 'Field of Dreams: Understanding Sociohistorical Changes in College Access.' In *Higher Education: Handbook of Theory and Research* William G. Tierney, ed. New York: Agathon Press.

Mead, George Herbert. 1934. *Mind, Self and Society.* Chicago: University of Chicago Press.

Merton, Robert K. 1938. *Science, Technology and Society in Seventeenth Century England.* Bruges: St Catherines Press.

Merton, Robert K. 1957. *Social Theory and Social Structure.* Toronto: The Free Press.

Merton, Robert K. 1968. 'The Matthew Effect in Science.' *Science* **159**: (January): 56–63.

Merton, Robert K. [1942] 1973. 'The Normative Structure of Science.' In *The Sociology of Science.* Chicago: University of Chicago Press.

Merton, Robert K. and Harriet Zuckerman. 1976. 'Age and Scientific Productivity.' In *The Sociology of Science.* Chicago: University of Chicago Press.

Miller, Jean Baker. 1976. *Toward A New Psychology of Women.* Boston: Beacon Press.

Moen, Phyllis. 1988. *Women as a Human Resource.* Washington DC: National Science Foundation, Sociology Program, Division of Social and Economic Science.

Moulton, R. 1972. 'Psychoanalytic Reflections on Women's Liberation.' *Contemporary Psychoanalysis* **8**: 197–233.

Moxham, H.J. Muir and L.A. Rogers. 1993. 'The Changing Place of Women in Academic Biomedicine in Britain.' In *The Position of Women in Scientific Research Within the European Community: Report of the Commission of the European Communities.* Talapessy, Lily, ed. Brussels: Directorate General for Science, Research and Development.

Moxham, H.J. Muir. 1993. 'Women in S/T Research in the United Kingdom.' In *The Position of Women in Scientific Research Within the European Community: Report of the Commission of the European Communities.* Talapessy, Lily, ed. Brussels: Directorate General for Science, Research and Development.

Murmann, Johann P. 1998. 'Knowledge and Competitive Advantage in the Synthetic Dye Industry, 1850-1914: The Coevolution of Firms, Technology, and National Institutions in Great Britain, Germany and the United States.' Unpublished doctoral dissertation, Columbia University.

Murmann, Johann P. and Ralph Landau. 1998. 'On the Making of Competitive Advantage: The Development of the Chemical Industries in Britain and Germany Since 1850.' In *Chemicals and Long-Term Economic Growth: Insights From the Chemical Industry.* A. Arora, R. Landau, and N. Rosenberg, eds. NY: John Wiley & Sons, Inc.

National Research Council. 1991. *Women in Science and Engineering: Increasing their numbers in the 1990's*. Washington DC: National Research Council Press.

National Resarch Council. 1994. *Women Scientists and Engineers Employed in Industry: Why So Few*. Washington DC: National Academy Press.

National Science Foundation. 1982. *Women and Minorities in Science and Engineering*. Washington DC: Government Printing Office.

National Science Foundation. 1984. *Women and Minorities in Science*.

National Science Foundation. 1988. *Science and Engineering Indicators*. Washington DC: Government Printing Office.

National Science Foundation. 1994. *Women, Minorities and Persons with Disabilities in Science and Engineering*. Washington DC: Government Printing Office.

National Science Foundation. 1996. *Science and Engineering Indicators*. Washington DC: Government Printing Office.

National Science Foundation. 1998. *Science and Engineering Indicators*. Washington DC: National Science Board.

National Science Foundation. 1999. *Women, Minorities and Persons with Disabilities in Science and Engineering*. Washington DC: Government Printing Office.

Nerad, Marisi. 1992. 'Using Time, Money and Human Resources Effectively in the Case of Women Graduate Students.' Paper prepared for the Conference Proceedings of 'Science and Engineering Programs: On Target for Women?' National Academy of Sciences.

Nuevo Kerr, L. A. 1993. 'Institutional and Demographic Frameworks for Affirmative Action in the 1990s.' *Journal of Women's History* **4**(3): pp.141–146.

Noble, David. 1992. *A World Without Women: the Christian Clerical Culture of Western Science*. Oxford: Oxford University Press.

O'Brien, P. C. 1988. 'Comparing Two Samples: Extensions of the T, Rank-Sum, and Log-Rank Tests.' *Journal of the American Statistical Association* **83**: 52–61.

Oleson, Alexandra and John Voss. 1979. *The Organization of Knowledge in Modern America: 1860–1920*. Baltimore: Johns Hopkins University.

Oncu, A. 1981. 'Turkish Women in the Professions: Why So Many?' In N. Abadan-Unat, ed. *Women in Turkish Society*. pp. 181–92. Leiden: E.J. Brill.

Orenstein, Peggy. 1994. *School Girls*. New York, NY: Doubleday.

Ortmeyer, Dale. 1988. *Psychoanalytic History and Issues of Gender*. Presented at William Alanson White Institute, New York City, January 12. [Cited in Friedman, Gloria and Barry Protter. 1995 'Gender and Sexuality.' *Handbook of Interpersonal Psychoanalysis*. Hillsdale, NJ: The Analytic Press.

Osborne, Mary. 1994. 'Status and Prospects of Women in Science in Europe.' *Science*, **263** pp. 1389–91

Palomba, Rosella. 1993. 'Women in S/T Research in Italy.' In *The Position of Women in Scientific Research Within the European Community: Report of the Commission of the European Communities*. Talapessy, Lily, ed. Brussels: Directorate General for Science, Research and Development.

Pearl, Amy *et al.* 1990. 'Becoming A Computer Scientist: A Report by the ACM Committee on the Status of Women in Computer Science. *Communications of the ACM*, November.

Pearson, Willie and Alan Fechter, eds. 1992. *Who Will Do Science?* Baltimore: Johns Hopkins University Press.

Pearson, Willie and Alan Fechter, eds. 1994. *Human Resources for Science*. Baltimore: Johns Hopkins University Press.

Perrow, Charles. 1986. *Complex Organizations: A Critical Essay, 3rd edition*. New York: Random House.

Pfeffer, Jeffrey. 1993. 'Barriers to the Advance of Organizational Science: Paradigm Development as a Dependent Variable.' *The Academy of Management Review* **18**(4): 599–620.

Pfeffer, Jeffrey and Jerry Ross. 1982. 'The Effects of Marriage and a Working Wife on Occupational and Wage Attainment.' *Administrative Science Quarterly* **27**: 66–80.

Podolny, Joel M. and James N. Baron. 1997. 'Resources and Relationships: Social Networks and Mobility in the Workplace.' *American Sociological Review* **62**: 673–93.

Porter, Beverly. 1989. 'Scientific Resources for the 1990's: Women the Untapped Pool.' *American Association for the Advancement of Science Annual Meetings.*

Powell, Walter W. and Jason Owen-Smith. 'Commercialism in Universities: Life Sciences Research and its Linkage with Industry.' *Journal of Policy Analysis and Management* **17**(2): 253–77.

Powell, Walter W., Kenneth W. Koput, and Laurel Smith-Doerr. 1996. 'Interorganizational Collaboration and the Locus of Innovation: Networks of Learning in Biotechnology.' *Administrative Science Quarterly* **41**: 116–45.

Powell, Walter W., Kenneth W. Koput, Laurel Smith-Doerr, and Jason Owen-Smith. 1998. 'Network Position and Firm Performance: Organizational Returns to Collaboration in the Biotechnology Industry.' In *Networks in and Around Organizations.* Steven Andrews and David Knoke, eds. Greenwich, CT: JAI Press.

Quinn, Susan. 1995. *Marie Curie.* New York: Simon and Schuster.

Radtke, Heidrun. 1991. 'Women in Science Careers in the German Democratic Republic.' In *Women in Science: Token Women or Gender Equality?* Veronica Stolte-Heiskanen, ed. Oxford: Berg Publishers.

Raffalli, Mary. 1994. 'Why So Few Women Physicists?' *New York Times* Education Life, p. 26, 28.

Rayman, Paula and Belle Brett. 1993. *Pathways for Women in the Sciences.* Wellesley, MA: Center for Research on Women.

Rayman, Paula, Cindy-Su Davis, Angela B. Ginorio and Carol S., Hellenshead. 1996. *The Equity Equation: Fostering the Advancement of Women in the Sciences, Mathematics, and Engineering* (Jossey-Bass Education Series, March 1996). San Francisco, CA: Jossey-Bass Publishers.

Reskin, Barbara. 1978. 'Sex Differentiation and the Social Organization of Science.' In *Sociology of Science.* Jerry Gaston, ed. San Francisco: Jossey Bass.

Rosenfeld, Rachel. 1984. 'Academic Career Mobility for Psychologists.' In *Women in Scientific and Engineering Professions.* Haas, Violet and Carolyn Perrucci (eds.) Ann Arbor: University of Michigan Press.

Rossi, Alice. 1965. 'Women In Science: Why So Few?' *Science* **148**: 1196–1203.

Rossiter, Margaret. 1978. 'Sexual Segregation in the Sciences: Some Data and a Model.' *Signs* **4**: 146–51.

Rossiter, Margaret. 1982. *Women Scientists in America.* Baltimore: Johns Hopkins University Press.

Ruivo, Beatriz. 1992. 'International Science Policy and Women: The Missing Issue.' Paper presented at European Association of Science and Technology Studies, Gothenburg, Sweden.

Ruivo, Beatriz. 1987. 'The Intellectual Labor Market in Developed and Developing Countries: Women's Representation in Scientific Research.' *International Journal of Scientific Education.*

Ruskai, M. B. 1990. *Worlds Women: Trends and Statistics 1970–1990.* United Nations, 1991.

Sadker, M. and D. Sadker. 1994. *Failing at Fairness: How America's Schools Cheat Girls.* New York: Charles Scribner's & Sons.

Sayre, Ann. 1975. *Rosalind Franklin and DNA.* New York: Norton

Schachtel, E. 1959. *Metamorphosis.* New York: Basic Books.

Schneer, J. A. and F. Reitman. 1993. 'Effects of Alternative Family Structures on Managerial Career Paths.' *Academy of Management Journal* **36**: 830–43.

Science 1993. Special Issue on Women in Science. Vol. **260**.

Scott, Joan. 1990. 'Disadvantage of Women by the Ordinary Processes of Science: the Case of Informal Collaboration'. In *Despite the Odds: Essays on Canadian Women and Science.* Marianne Ainley, ed. Montreal: Vehicule Press.

Seashore, Karen, David Blumenthal, Michael Gluck, and Michael Soto. 1989. 'Entrepreneurs in Academe: An Exploration of Behavior Among Life Scientists.' *Administrative Science Quarterly* **34**: 110–131.

Seymour, Elaine. 1995. 'The Loss of Women from Science, Mathematics and Engineering Undergraduate Majors: An Explanatory Account.' *Science Education* **79**(4): 437–73.

Seymour, Elaine and Nancy M. Hewitt. 1997. *Talking About Leaving: Why Undergraduates Leave the Sciences.* Boulder, CO: Westview Press.

Shapiro, Lucy and Susan Henry. *The Maverick and the Maize: The Work and World of Barbara McClintock.*

She, Hsiao Ching. 1995. 'Elementary and Middle School Students' Image of Science and Scientists Related to Current Science Textbooks in Taiwan.' *Journal of Science Education and Technology* **4**(4): 283–94.

Sime, Ruth. 1996. *Lise Meitner.* Berkeley: University of California Press

Simon, Herbert A. 1963. 'Economics and psychology.' In *Psychology: A Study of Science.* S. Koch. ed. Vol. 6. New York: McGraw-Hill.

Sjoberg, Sven. 1988. 'Gender and the Image of Science.' *Scandinavian Journal of Educational Research*, pp. 49–60.

Sonnert, Gerhard. 1996. 'Gender Equity in Science: Still An Elusive Goal.' *Issues in Science and Technology* (Winter): 53–8.

Sonnert, Gerhard and Gerald Holton. 1994. *Gender Differences in Science Careers.* New Brunswick: Rutgers University Press.

Sonnert, Gerhard and Gerald Holton. 1996. 'Career Patterns of Women and Men in the Sciences.' *American Scientist* **84**: 63–71.

South, S. J., C.M. Bonjean, W.T. Markham, and J. Corder. 1982. 'Social Structure and Intergroup Interaction.' *American Sociological Review* **47**: 687–599.

Spangler, E., M.A. Gordon and R. M. Popkin. 1978. 'Token Women.' *American Journal of Sociology* **85**: 160–70.

Spertus, Ellen. 1991. 'Why Are There So Few Female Computer Scientists?' Cambridge M.I.T. Artifical Intelligence Laboratory. Technical Report 1315.

Stata. 1996. *Stata Release 5.* College Station Texas: Stata Press.

Stolte-Heiskanen, Veronica. 1983. 'The Role and Status of Women Scientific Research Workers in Research Groups.' In *Research in the Interweave of Social Rules: Jobs and Families* 3: 59–87. Greenwich CT: Jai Press.

Stolte-Heiskanen, Veronica. 1987. 'Women and Science: The Role of Gender Relations in Knowledge Production.' *Current Sociological Perspectives* **6**(1&2) (Spring and Autumn): 121–42.

Stolte-Heiskanen, Veronica, ed. 1991. *Women in Science: Token Women or Gender Equality?* Oxford: Berg Publishers.

Stroh, Linda K. and Jeanne M. Brett. 1996. 'The Dual-Earner Dad Penalty in Salary Progression.' *Human Resource Management* **35**: 181–201.

Tabak, Fanny. 1993. 'Women Scientists in Brazil: Overcoming National, Social and Professional Obstacles.' Paper presented at the Third World Organization of Women Scientists, Cairo, January.

Talapessy, Lily, ed. 1994. *The Position of Women in Scientific Research Within the European Community: Report of the Commission of the European Communities.* Brussels: Directorate General for Science, Research and Development.

Tobias, Sheila and Frans Birer. 1998. 'The Science-Trained Professional: A New Breed for the New Century.' *Industry and Higher Education* **12**(4): 213–16.

Toren, Nina. 1991. 'The Nexus Between Family and Work Roles of Academic Women in Israel: Reality and Representation. *Sex Roles* **24**(11/12): 651–67.

Toren, Nina and Vered Kraus. 1987. 'The Effects of Minority Size on Women's Position in Academia'. *Social Forces* **65**(4): 1090–1100.

Traweek, Sharon. 1988. *Beamtimes and Lifetimes.* Berkeley: University of California Press

Tronick, Edward and Lauren Adamson. 1980. *New Findings on our Social Beginnings.* New York: MacMillan.

Ullman, Montegue. 1975. 'The Transformation Process in Dreams.' *The Academy: Journal of the American Academy of Psychoanalysis* **19**: 2.

Ullman, Montegue. 1992. 'An Approach to Closeness: Dream Sharing in a Small Group.' In *Closeness.* Boston: Shamchala Press.

Uzzi, Brian. 1996. 'The Sources and Consequences of Embeddedness for the Economic Performance of Organizations.' *American Sociological Review* **61**: 674–98.

Uzzi, Brian. 1997. 'Social Structure and Competition in Interfirm Networks: The Paradox of Embeddedness.' *Administrative Science Quarterly* **42**: 35–67.

Uzzi, Brian. 1999. 'Embeddedness in the Making of Financial Capital. How Social Relations and Networks Benefit Firms Seeking Capital.' *American Sociological Review* **64**: 181–205.

Uzzi, Brian and Zoe Barsness. 1998. 'Inside Out Employment Arrangements in British Firms: The Structural and Organizational Determinants of the Use of Externalized Workers.' *Social Forces* **76**, Number 3.

Vazquez, Analia. 1993. 'La Situation De La Mujer En La Universidad, En El Mercado Laboral Y En La Investigacion En La Argentina'. Paper presented at the Third World Organization of Women Scientists, Cairo, January.

Vetter, Betty. 1992. In *Who Will Do Science?* Willie Pearson and Alan Fechter, eds. Baltimore: Johns Hopkins University Press.

Watson, James. 1968. *The Double Helix.* New York: Norton.

Watts, Duncan J. and Steven H. Strogatz. 1998. 'Collective Dynamics of 'Small-World' Networks.' *Nature* **393**: 440–22.

The Wellesley College Center for Research on Women. 1992. *How Schools Shortchange Girls: The AAUW Report.* Washington, DC: American Association of University Women Educational Foundation.

White, Harrison C. 1992. *Identity and Control: A Structural Theory of Action.* Princeton, NJ: Princeton University Press.

White, Robert W. 1974. 'Strategies of Adaptation: An Attempt at Systematic Description'. In *Coping and Adaptation.* George V. Coelho, David A. Hamburg and John E. Adams, eds. New York: Basic Books.

Will, J.A., P.A. Self and N. Datan. 1976. 'Maternal Behavior and Perceived Sex of Infant.' *American Journal of Orthopsychiatry* **46**: 135–9.

Zuckerman, Harriet and Robert K. Merton. 1972. 'Age, Aging, and Age Structure in Science.' In *Aging and Society: A Sociology of Age Stratification.* M. W. Riley, M. Johnson, and A. Foner, eds. Vol. 3. New York: Russell Sage Foundation.

Zuckerman, Harriet, Jonathan Cole, and John Bruer. 1991. *The Outer Circle: Women in The Scientific Community.* New York: Norton.

Index

Key to index: n following page number indicates information in a footnote; Ap indicates information in the Appendix. Page numbers in bold refer to tables.

closest to a gender-neutral element 76
effects on women 75–6
qualities, of femaleness and maleness, not
    rigid 46
quota hires 228,229

'Re-entry' program, University of California,
    Berkeley
    funded outside the department 192–3
    success of 190
reciprocity/reciprocation 160
    between contacts 129
    lower levels of for women 161–2
relational departments 181–3,200
    attraction of interpersonal interactions 181
    effects of cooperative atmosphere 181
    importance of sympathetic leadership 182
    improved quality of life for women 200
relational style 148,153–5
    emphasis on collaboration and community
        154
research
    Austria, women in positions of importance
        helpful 212–13
    importance of bridging ties 169
    lab relationships important for strategy
        154
    Netherlands 216
    in out-of-the-way fields 130
    productivity and social capital 173–6
    success of today's projects 120–1
    in the U.S. departmental model 71
    women participate most in areas of fastest
        growth 212
    women in, Spain 220
    women's, presentations in safe
        environments 191,192
risk taking, women lack support for 86
role models 13–14,105
    complexities of being 148
    felt to be lacking, female Ph.D. students 87
    for role models 247
    stressing positive or negative sides? 13
    through removal of barriers 237,238
Rossi, Alice 1
    on marriage, parenthood and career 247
    'technical fix' alternative 216
    on young girls with high ability 44

Science: The Endless Frontier, Vannevar Bush
    118–19

Science, article, biological male superiority
    and standardized testing 45–6
science
    academic, negative female image of 137
    conditions for successful career in 124
    continual departure of girls and women
        from 155
    covert resistance to women persists 221–2
    dual male and female worlds 137–46
        resistance to change 142–5
        tenure 141–2
        tenure stress 145–6
    emergence of female-gendered subfields
        112–13
    foreclosing on women's choice to do 47–8
    gender inequality and shortage 229–31
    girls' interest discouraged during
        adolescence 42–7
    graduate experience in 83–103
    hampered by long-term relative exclusion
        of women 25
    high-level careers for women to be seen as
        normal 237–8
    interpersonal networks differ for men and
        women 17
    lagging in its inclusion of women 2
    low status, aids women's participation
        203,205–6
    male culture makes women invisible 99
    non-sexist framework, incorporating male
        and female perspectives 249–50
    open to all talent, an emerging picture 241
    paradox of women's participation 203–4
    permeated by male standards of behavior 26
    personal qualities needed for success
        changing 26
    relationship to other spheres of life should
        be rethought 247
    seen as 'masculine' 31,32
    sociology of is moving on 241–2
    stereotyping of in the primary school years
        38–42
        boys, use of negative/inappropriate
            behavior 40
        compliance by girls costly 40
        enlightened parents dismayed at sexual
            stereotyping 39
        girls, teachers less responsive to 39–40
        girls tend to avoid lack of structure 41
        masculine image already established
            38–9